# CONTEMPORARY TRENDS IN ADHD RESEARCH

Edited by **Jill M. Norvilitis**

**Contemporary Trends in ADHD Research**

Edited by Jill M. Norvilitis

**Published by InTech**

Janeza Trdine 9, 51000 Rijeka, Croatia



**Publishing Process Manager** Igor Babic

**Technical Editor** Teodora Smiljanic

**Cover Designe**r InTech Design Team

Additional hard copies can be obtained from orders@intechopen.com

Contemporary Trends in ADHD Research, Edited by Jill M. Norvilitis
p. cm.
ISBN 978-953-307-858-8

For EU product safety concerns:
IN TECH d.o.o., Prolaz Marije Krucifikse Kozulić 3, 51000 Rijeka, Croatia,
info@intechopen.com or visit our website at intechopen.com.

# Meet the editor

Dr Jill M. Norvilitis PhD is a Professor of Psychology at Buffalo State College in Buffalo, NY. She earned her undergraduate degree in psychology from Hope College in Holland, Michigan and her MA and PhD in Clinical Psychology from Wayne State University in Detroit, Michigan. In 2010, she was recognized with the SUNY Chancellor's Award for Excellence in Teaching. Dr. Norvilitis is a Clinical Psychologist who specializes in behavior disorders of childhood. Her research focuses on predictors of social and academic adjustment among college students with Attention Deficit Hyperactivity Disorder (ADHD) and perceptions of ADHD cross-culturally. She is the author of over 30 journal articles, book chapters, and edited volumes and she has given over 80 conference and invited presentations.

# Contents

# Preface

Attention Deficit Hyperactivity Disorder is one of the most widely diagnosed behavior disorders in childhood, with a worldwide pooled prevalence of 5.29 % (Polanczyk et al., 2007). With so many children and adults affected by the disorder, researchers strive on multiple fronts to understand the underpinnings of ADHD and associated factors. At a basic level, our understanding of the biological origins is not definitive and much remains to be learned. While scientists work to unlock these answers, others work to improve the lives of children with ADHD as the children struggle academically and socially. Equally vital, this applied work addresses many of the immediate needs and concerns of this population.

This goal of this book is to explore some of the broad array of research in the field of ADHD. The 12 chapters represent the work of 23 researchers from countries around the globe and cover such topics as varied as postural control, endocrine dysfunction, juvenile justice, and academic outcomes.

The first section of the book explores basic research in the field, examining questions related to the nature of the disorder. Chapter 1, "Sensory integration in attention deficit hyperactivity disorder: Implications for postural control," investigates difficulties in static postural control in children with the combined subtype of ADHD. Results indicate significantly increased risk for poor balance, particularly in challenging situations, suggesting the need to evaluate postural control in children referred for ADHD and to consider the role of such deficits in the reported poor coordination and clumsiness of children with the disorder. Chapter 2, "Variability, noise and predictability in motor response times: Adaptation or misadaptation?" considers intra-individual-variability in cognitive performance. Although such variability is often treated as error or noise in testing, the authors present compelling evidence for the utility of analyzing and understanding such variability. Chapter 3, "Endocrine dysfunction and growth in attention deficit hyperactivity disorder," reviews the conflicting literature on hypothalamic-pituitary-adrenocortical axis activity in ADHD, and also examines growth and growth hormone treatment in children with ADHD.

Chapter 4, "Attention deficit hyperactivity disorder: Birth season and epidemiology," describes the prior work on the relationship between season and month of birth, and

ADHD, which has found conflicting results, with differing studies indicating higher rates of ADHD birth in autumn, spring and summer and still other research reporting no relationship at all. The authors also report the results of their own research, which did not support a relationship between birth season and ADHD. Chapter 5, "Gestational hypothyroxinemia is associated with attention deficit hyperactivity disorder in the progeny of Spanish mothers," examines the role of prenatal hypothyroxinemia in ADHD through a longitudinal study of children born to mothers with and without thyroid problems during pregnancy. Although there were no other differences at birth, at the time of assessment, the children exposed to low prenatal thyroid hormones were at significantly increased risk for inattention and hyperactive symptoms. These intriguing results suggest a need for improved follow up for children exposed to hypothyroxinemia. Chapter 6, "Cutting corners: Neuropsychological research into the energetics of ADHD," explores the Cognitive Energetic Model of ADHD. The author discusses bottom up motivational processes and effort allocation in the combined and inattentive subtypes of the disorder, noting significant differences between the two subtypes.

The second half of the book addresses ADHD in applied settings, such as school, treatment, and the juvenile justice system. Chapter 7, "Adolescent academic outcome of childhood attention deficit hyperactivity disorder—a population based study," examines the role of ADHD symptoms in the academic achievement of both clinically diagnosed and subthreshold children. The results indicate significant academic risks for both groups and highlight the need for multi-disciplinary intervention.

Chapter 8, "Remedial education for children with ADHD in Sweden," continues this vein and addresses educational issues for children with ADHD in general and, more specifically, within the Swedish educational system, highlighting the critical need for teacher training and proactive interventions. Chapter 9, "Evaluation of the level of knowledge of infant and primary school teachers with respect to attention deficit hyperactivity disorder (ADHD): Content validity of a newly created questionnaire," addresses the need to be able to assess teachers' knowledge of and attitudes toward ADHD. The authors describe the creation of a new standardized measure that will allow for the assessment of the efficacy of teacher training and to examine to what degree teacher knowledge of ADHD affects student outcomes.

Chapter 10, "ADHD symptomatology, academic dishonesty, and the use of stimulant medications without a prescription," examines the increasing misuse of stimulant medications for ADHD among college students by those who do not have prescriptions. Such users are more likely to have more symptoms of undiagnosed ADHD and to struggle with their motivation to attend college, suggesting a need for both prevention and intervention. Chapter 11, "Understanding the distracted and the disinhibited: Experiences of adolescents diagnosed with ADHD with the African context," gives a voice to adolescents in treatment for ADHD in South Africa through qualitative analysis. The participants candidly discuss their problems, treatment and self-perceptions. Chapter 12, "Attention deficit hyperactivity disorder and males in the juvenile justice system," provides an overview of the prevalence of ADHD in the juvenile justice system and the need for screening for the disorder to help reduce recidivism.

Overall, the chapters presented here tap much of the breadth of the field. I believe that these chapters represent the state-of-the-art in ADHD research and it is my hope that they will provide valuable insights for students reading about ADHD for the first time, researchers wishing to learn about the latest advances, and practitioners seeking new insight in the field.

This book is the result of the work of many individuals. I am particularly grateful for the assistance of Ms. Adriana Pecar and Mr. Igor Babic for their assistance in coordinating this book. I also thank all of the authors who contributed to this volume: I have learned from you and been inspired by your work.

## References

Polanczyk, G., Silva de Lima, M., Lessa Horta, B., Biederman, J., & Rohde, L. A. (2007). The worldwide prevalence of ADHD: A systematic review and meta-regression analysis. *American Journal of Psychiatry, 164, 6*, 942-948.

**Jill M. Norvilitis**
Department of Psychology,
Buffalo State College,
Buffalo, NY
USA

# Part 1

## Basic Research in ADHD

# 1

# Sensory Integration in Attention Deficit Hyperactivity Disorder: Implications to Postural Control

Dalia Mohamed Hassan[1] and Hanan Azzam[2]
*[1]Audiology Unit, ORL Department,*
*[2]Department of Neuro-Psychiatry,*
*Faculty of Medicine, Ain Shams University, Cairo,*
*Egypt*

## 1. Introduction

A major task of the central nervous system is to configure the way in which sensory information becomes linked to adaptive responses and meaningful experiences. The neural systems that bridge the gap between sensation and action provide the substrates for 'intermediary' or 'integrative' processing (Miller et al., 2009). Sensory integration disorder 'SID' is a neurological disorder that results from the brain's inability to integrate certain information received from the body's five basic sensory systems (vision, auditory, touch, olfaction, and taste), the sense of movement (vestibular), and/or the positional sense (proprioception). Sensory information is sensed normally, but perceived abnormally affecting participation in functional daily life routines and activities (Bundy et al., 2002).

Around 16 percent of the general population has symptoms of SID. In attention deficit hyperactivity disorder 'ADHD', the frequency of SID rises to 40 - 84% as reported in different studies (Mulligan, 1996; Dunn & Bennett, 2002; Ben-Sasson et al., 2009). One of the categories proposed within SID included sensory-based motor disorder. Sensory-based motor disorder comprises postural disorder (which reflects problems in balance and core stability) and dyspraxia (which encompasses difficulties in motor planning and sequencing movements) (Miller et al.,2007, 2009; Buderath et al., 2009).

Static postural control (stability) is the ability to maintain center of mass (center of gravity) within the base of support (Horak, 1987). The integration of the sensory information from somatosensory, visual, and vestibular origins by the central nervous system, followed by coordinated automatic outputs involving the muscles of postural control is crucial to maintain stability and orientation of the body to the environment (Hunter & Hoffman, 2001). With children, postural stability is gradually acquired as various systems mature, greater experiences accumulate and sensory integration takes place. They begin to approximate adult levels of performance by the age of seven years (Palmeri et al., 2002; Shepard & Janky, 2008).

What is not understood is the developmental profile of children with ADHD. Children with ADHD have been found to have an increased velocity of postural sway than normal children (Zang et al., 2002; Shum & Pang, 2009). In daily activities, they manifest problems performing certain athletic sports, were frequently and involuntarily bumping into things, lacking bounce when walking and running, and became more easily tired and exhausted than peers (Stray etal., 2009).

Computerized Dynamic Posturography 'CDP' assesses the functional capacity of the balance system in an objective and quantifiable manner. By systematically manipulating support surface and visual surround, the sensory organization test (SOT) is an important tool which helps quantify the sensory contributions that aide in sensory integration and the development of postural control (Shepard & Telian, 1996). It evaluates the ability to use in combination or individually the three sensory inputs during maintenance of stance. Information about the automatic patients' reactions to unexpected external disturbances in their centre of mass position is obtained from the motor control (MCT). Furthermore, the adaptation (ADT) test illustrates the response adaptation to irregular/varying support surface conditions. Both MCT and ADT evaluate the postural control long loop pathway (Allum & Shepard, 1999).

Balance deficits are usually not addressed with ADHD children because awkwardness and clumsiness are likely attributed to lack of "attention or concentration". This study was designed to compare the static postural control function in a group of ADHD/C children and typically developing (TD) children using CDP. This might be considered as a step to investigate one of SID subtypes in the studied children.

## 2. Methodology

### 2.1 Patients

Twenty children with ADHD of the combined subtype (ADHD/C) were included in the present study. They were diagnosed according to the diagnostic and statistical manual 'DSM-IV' criteria for ADHD (American Psychiatric Association, 1994). Selection of children was randomly obtained from the clinic records of the psychiatry outpatient clinic, Institute of Psychiatry, Ain Shams University Hospitals during the period from January 2010 to July 2010. Informed consent was taken from the parents with explanation of the test procedures, benefits, and risks according to the ethical rules.

Selection of children considered an age range between eight and ten years. Intelligent Quotient (IQ) should be more than 85 using Wechsler Intelligence Test for Children 'Arabic version'. A minimum score of 70 (markedly atypical) on at least 2 subscales of the Conner's Parent Rating scale was an important inclusion criterion. Children should be free from neurological, sensory, and orthopaedic problems and not on psychotropic medications.

Twenty age, sex and height matched typically developing (TD) children were used as a control group. They had no history suggestive of behavioral, attention problems, medical, hearing, balance, orthopaedic, visual or neurological disorders.

### 2.2 Procedures

Careful history taking and neuro-psychiatric assessment was performed by a child psychiatrist. The Arabic version of the Mini-International Neuro-psychiatric Interview for Children and Adolescents (M.I.N.I-Kid) was applied to confirm the ADHD diagnosis, subtype and exclude other co-morbid conditions. MINI-Kid is a short, structured interview designed to assess symptoms of several Axis I disorders as listed in the DSM-IV and the International Statistical Classification of Diseases and Related Health Problems (Ismail & Melika, 1961). Assessment of IQ was done using Wechsler intelligence scale for children (Sheehan et al., 1998) by a clinical psychologist.

To assess the degree of ADHD severity, the Conner's parent rating scale revised, long version (CPRS-R-L) was used (Conner, 1997). It represented an 80 items questionnaire with an average administration time of 25-30 minutes. It scored the parents report of their child's behavior during the past month on a 4-point response scoring.

In the vestibular clinic, Ain Shams university hospitals, the postural control system was tested for all children by an audiologist. It was done using Computerized Dynamic Posturography 'CDP' SMART EquiTest system. The CDP sub-tests used were: sensory organization test 'SOT', motor control test 'MCT', and adaptation test 'ADT'. The test procedure, instructions, and analysis followed the SMART EquiTest system manual version 8 specifications.

The SOT measured the ability to perform volitional quiet stance during manipulation of the different sensory inputs available for use. During the SOT, the somato-sensory and visual environments were altered systematically through movement of forceplate, visual surround, or both. Six conditions of the SOT assessment were applied as illustrated in (Figure 1). The system recorded data for a maximum of three trials for each of the six conditions. Each trial lasted 20 seconds. Prior to each trial the child was given the proper instructions.

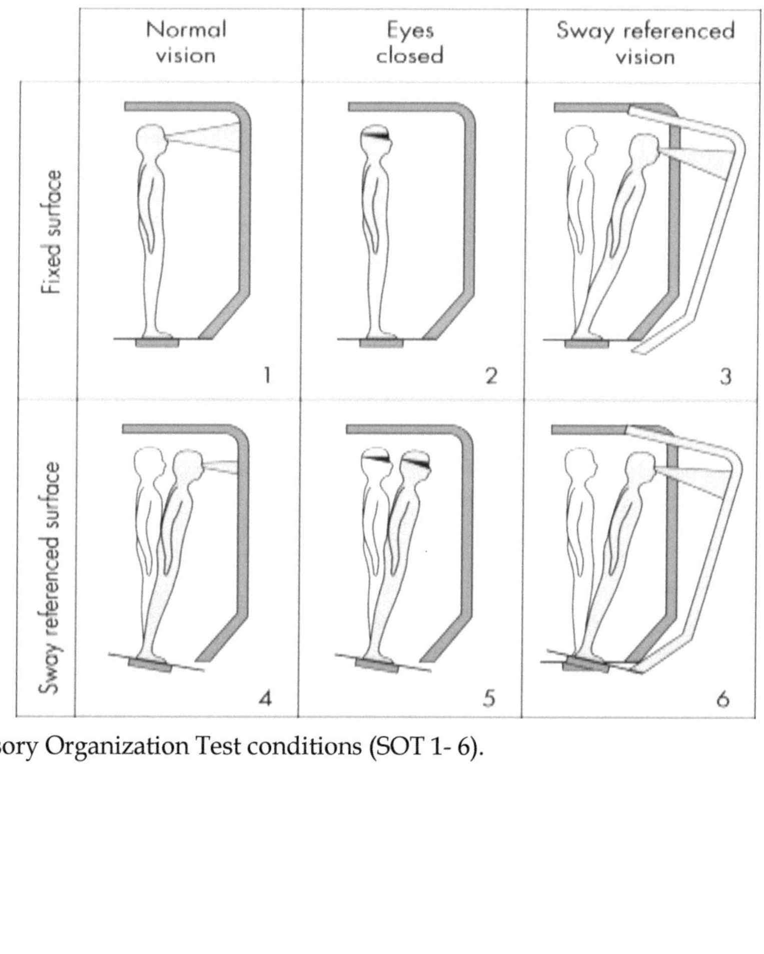


Fig. 1. Sensory Organization Test conditions (SOT 1- 6).

The data obtained from SOT analysis were:

- Equilibrium Score: It is a percentage score reflecting the magnitude of sway of centre of mass in the sagittal plane for each trial of the 6 sensory conditions. The normal value of patient's sway limit should be within 12.5 degrees of sway in the antero/posterior direction, 8 and 4.5 degrees in forward and backwards directions, respectively. A patient swaying to these limits will receive a very low score. The highest possible score was 100, which indicates that the patient did not sway at all. The composite equilibrium score was also recorded.
- Sensory Analysis: It included the sensory ratios computed from the average equilibrium scores obtained on specific pairs of sensory test conditions as described in table 1.

| Sensory ratio | SOT conditions | Significance |
|---|---|---|
| Somatosensory 'SOM' | Condition 2/condition 1 | Patient's ability to use input from the Somatosensory system to maintain balance |
| Visual 'VIS' | Condition 4/condition 1 | Patient's ability to use input from the visual system to maintain balance |
| Vestibular 'VEST' | Condition 5/condition 1 | Patient's ability to use input from the vestibular system to maintain balance |
| Visual preference 'PREF' | Condition3+6 /condition 2+5 | Degree to which patient relies on visual information to maintain balance, even when the information is incorrect |

Table 1. Computation of the sensory analysis ratios

- Strategy Analysis: It showed the relative amounts of movement about the ankles (ankle strategy) and about the hips (hip strategy) that the patient used to maintain balance during each procedure. Exclusive use of ankle strategy to maintain equilibrium resulted in a score of 100. Exclusive use of hip strategy would give a score near 0. Scores between these two extremes represented a combination of the two strategies.

The *MCT* assessed the ability of the automatic motor system to quickly recover following an unexpected external disturbance. This demonstrated the patient's ability to coordinate automatic movement responses to maintain standing posture. Three sequences of platform translations of varied sizes (Small, medium and large) were administered in forward and backward directions lasting less than one second. The sizes of the translations were scaled to the patient's height to produce sway disturbances of equal size. A random delay of 1.5 to 2.5 seconds was between the trials. For the child to perform the test, weight-bearing symmetry was ensured to be within the normal limits.

The Measurements collected from the MCT were the speed of reaction (latency), and the relative response strength. The *Latency* was defined as the time in milliseconds (ms) between the onset of a translation and the onset of the patient's active response to the support surface movement. The *relative response strength* was calculated as the amplitudes of the patient's active response to each size and direction of translation in degrees/sec. Values for each leg in the small, medium and large movements and in the forward and backward direction were also obtained.

The *ADT* demonstrated the ability of the automatic postural control to adapt to recurrent surface movements. A series of rotary platform movements, making the patient's toes to go up or down, were used. Rotations lasted 0.4 seconds and with uniform amplitude for all trials (8°). There were five trials for each type of rotation with a random delay of 3.0 to 5.0 seconds. The reaction force generated by the patient to minimize AP sway was measured.

Initially, the TD children group was tested to obtain norms for the 8-10 years age group. These normative data were subsequently used for comparison with the results obtained from ADHD/C children. To maximize subject familiarity with the tests, subjects practiced each assessment exercise before data collection. Subjects performed without shoes and socks. A harness was loosely fastened around the participant to prevent the participant from falling.

**Statistical analysis:** Statistical analyses were performed using (SPSS) 10.1. The Student's t test was used to analyze differences between the study groups. For comparing the variables in each group, the paired t test was applied. A level of $p < 0.05$ was considered significant while $p < 0.01$ was highly significant. A statistician was used for guidance in the study.

## 3. Results

Both ADHD/C and TD children were age and sex matched. They had mean age 8.9 (Standard Deviation 'SD' 0.9) and 9.2 (± 0.8) years, respectively. The ADHD/C group included 16 males and 4 females while the TD had 15 males and 5 females. The Conner's parent rating scale revised showed mean ADHD index scores = 73, mean clinical global impression for restless and impulsive = 79, mean total clinical global impression = 81. All these values reflected the severity of the ADHD condition. According to the parents' reports, four of the ADHD/C children frequently fall during running and three children had difficulty to engage in the gym class at school.

Looking to the CDP test results, the TD children group had mean values that approached the adult values (in the age range 20-59 years) in nearly all tests. On the other hand, children with ADHD had statistically significant lower mean SOT equilibrium scores in the six tested conditions and lower mean equilibrium composite score ($p < 0.05$). More difficulty was encountered in SOT conditions 5 and 6. The lowest scores and the greater difference in scores between the two groups were obtained in these two challenging conditions (Table 2). The SOT test was interrupted in five ADHD/C children as they tended to fall (three children in condition 6 and two children in condition 5 & 6).

| Group | SOT1 | | SOT2 | | SOT3 | | SOT4 | | SOT5 | | SOT6 | | Comp | |
|---|---|---|---|---|---|---|---|---|---|---|---|---|---|---|
| | X | SD | X | SD | X | SD | X | SD | X | SD | X | SD | X | SD |
| **ADHD** | 89 | 4 | 85 | 4 | 84 | 5 | 71 | 9 | 55 | 11 | 31 | 21 | 58 | 22 |
| **TD** | 93 | 3 | 90 | 3 | 88 | 5 | 81 | 6 | 70 | 12 | 61 | 9 | 73 | 7 |
| **t value** | -2.8 | | - 4 | | - 1.9 | | - 3.9 | | - 3.7 | | - 5 | | - 2.5 | |
| **p value** | 0.01* | | 0.001* | | 0.04* | | 0.001* | | 0.001* | | 0.001* | | 0.02* | |

$p < 0.05$ = statistically significant. Comp = composite equilibrium score.

Table 2. The equilibrium scores (%) obtained in the different SOT conditions in both study groups.

The sensory analysis showed that ADHD/C had lower somatosensory, visual, vestibular ratios by 1%%, 9%, and 18%, respectively compared to the TD children (Figure2). This difference was statistically significant for the visual and vestibular inputs ($p < 0.05$).

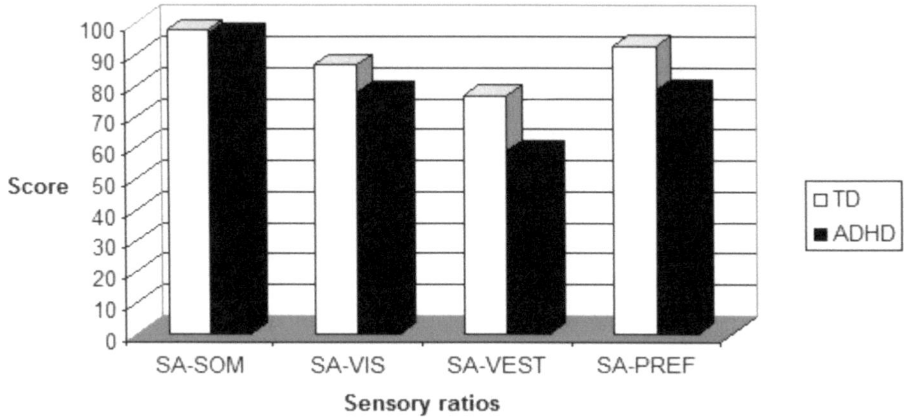


SOM: Somatosensory, VIS: Visual, VEST: Vestibular, PREF: Preference

Fig. 2. Sensory analysis (SA) ratios in both study groups.

Both groups used predominantly the ankle strategy during all SOT conditions to maintain equilibrium with no statistical significant difference detected. The strategy score in SOT conditions 1 - 6 was 98 (± 0.6), 98 (± 1.2), 97.5 (± 2), 87 (± 5.2), 80 (± 6), and 71 (± 8) respectively in ADHD children. In the TD children, it was 99 (±1.7), 98 (±1.6), 97 (± 2), 89 (± 5), 88 (± 7), and 74 (± 9) respectively.

In the MCT, prolonged latencies were observed in ADHD/C children relative to the TD group. The difference between the two groups reached statistical significance in more than one test condition ($p < 0.05$) (Table 3a, 3b). Both groups demonstrated comparable relative response strength. The right / left leg responses in each group did not show statistical significant difference in all test conditions.

The ADT scores were higher in the ADHD/C children in the two test situations (toes up & down) when compared to the TD children. This difference was statistically significant. The TD children had values approaching the adult values that decreased with increase the trial number. In ADHD/C children, the scores did not differ among the five conditions (Fig. 3a,b).

| **Movement Group** | **Small L** | | **Medium L** | | **Large L** | | **Small R** | | **Medium R** | | **Large R** | |
|---|---|---|---|---|---|---|---|---|---|---|---|---|
| | X | SD | X | SD | X | SD | X | SD | X | SD | X | SD |
| **ADHD** | 120 | 11 | 123 | 17 | 123 | 11 | 122 | 10 | 123 | 7 | 129 | 25 |
| **TD** | 113 | 31 | 122 | 9 | 117 | 11 | 109 | 42 | 123 | 9 | 117 | 9 |
| **t value** | - 0.8 | | 0.2 | | 1.5 | | - 1.2 | | - 0.1 | | 1.7 | |
| **p value** | 0.2 | | 0.4 | | 0.05 | | 0.1 | | 0.4 | | 0.04* | |

L = left leg, R = right leg, p < 0.05 = statistically significant.

Table 3a. The MCT latency in both groups in each leg during backward movements.

| Movement Group | Small L | | Medium L | | Large L | | Small R | | Medium R | | Large R | |
|---|---|---|---|---|---|---|---|---|---|---|---|---|
| | X | SD | X | SD | X | SD | X | SD | X | SD | X | SD |
| ADHD | 143 | 31 | 139 | 26 | 140 | 27 | 148 | 32 | 134 | 22 | 142 | 29 |
| TD | 129 | 13 | 125 | 9 | 126 | 13 | 135 | 31 | 125 | 10 | 126 | 17 |
| t value | 1.6 | | 2 | | 1.8 | | 1.2 | | 1.4 | | 2 | |
| p value | 0.06 | | 0.02* | | 0.03* | | 0.1 | | 0.08 | | 0.03* | |

L = left leg, R = right leg, $p < 0.05$ = statistically significant.

Table 3b. The MCT latency in both groups in each leg during forward movements.

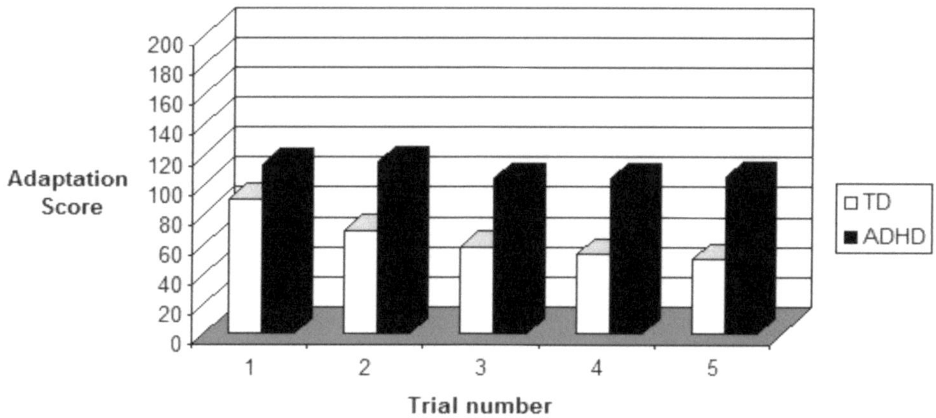


Fig. 3a. Adaptation test results toes up condition in both study groups.

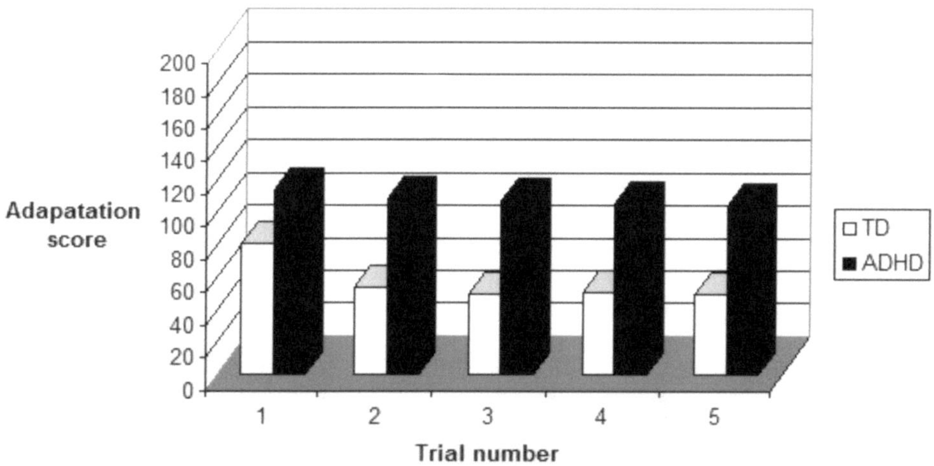


Fig. 3b. Adaptation test results toes down condition in both study groups

## 4. Discussion

In the present study, children with ADHD/C could not maintain quiet stance and showed more sway while performing all SOT conditions. The composite equilibrium score was 15% lower than the TD children (table 1). This could be the result of a lack of adequate interaction among the three sensory inputs that provide orientation information to the postural control system (Guskiewicz & Perrin, 1996). Higher equilibrium scores in the TD children indicated better coping mechanisms to balance perturbations (Bauer et al., 2001).

Poor stability with significant deficits in SOT was reported in ADHD/C by Shum & (Pang (2009) and Cherng et al (2001). As the individual matures and develops, sensory integration mechanisms are developed to suppress or inhibit irrelevant information and keep an excess of central nervous system arousal in check (Wang et al., 2003). This particular feature of development appears to be absent in individuals with ADHD. A lack of inhibition and sensory-motor homeostasis is linked to a lazy frontal lobe with the ADHD population and inadequate vestibular and somato-sensory feedback (Ayers, 1972; Mulligan, 1996; Zang et al., 2002).

Notably in this work, difficulties in postural control in ADHD/C showed up more clearly in the greater task constraints, evidenced by lower equilibrium scores in SOT conditions 5 and 6 with a tendency to fall in five children (25%). From SOT and sensory analysis, the vestibular system appeared to be less than fully developed sensory system relative to the somatosensory and visual systems. ADHD/C could not depend solely on the vestibular system information, resulting in poor scores in SOT conditions 5 & 6. In these conditions, the vestibular system is the only accurate system contributing to posture control (Shepard & Telian, 1996).

The vestibular system is known be less than adequate in individuals diagnosed with ADHD as reported by Zang et al (2002). They found that ADHD children were more dependent on visual feedback during the execution of the movement. It is well known that of the three sensory systems, the vestibular apparatus is the one lagging behind in development (Cherng et al., 2001). This phenomenon was more pronounced in the studied ADHD/C when compared to the TD children suggesting a delay in the maturation process that involves the vestibular system. An intact vestibular system is crucial to normal levels of arousal, attention and motor planning (Mulligan, 1996).

Furthermore, children with ADHD/C needed more time to recover from the unexpected disturbances in the support surface compared to the TD children. Prolonged latencies are strong evidence of musculoskeletal/biomechanical problems and/or pathology within the long loop pathways including the peripheral nerves, ascending and descending spinal pathways, and brain structures involving brainstem, basal ganglion, cerebellum and motor cortex (Shepard & Telian, 1996).

Although exposed to destabilizing rotary stimuli in the ADT, the TD children showed an appropriate corrective response to prevent fall after the first trial. Sway responses to the first rotation were typically larger than those of subsequent rotation, because patients usually reduce the resistance of their ankle joints to subsequent rotations. A normal postural control system is able to modify its response as an adaptive learning system (Shepard & Janky,

2008). On the other hand, the ADHD/C children generated more force than the normal children to minimize the antero-posterior sway ($p < 0.05$). They could not adapt to the randomly presented familiar destabilizing rotations on repeated trials (Figure 3a, 3b). Hence, a difficulty in motor learning and adaptation to change was suspected in those children.

Altered brain activity in children with ADHD could explain the sensori-motor deficits seen in the MCT and ADT in this study. The possible involved brain areas are the right inferior frontal cortex, left sensorimotor cortex, basal ganglia, and bilateral cerebellum and the vermis as well as in the right anterior cingulated cortex, and bilateral brainstem (Niedermeyer & Naidu, 1997). Numerous MRI studies observed smaller cerebellar volume with a particular reduction in the posterior inferior vermis in ADHD children (Bledsoe et al., 2009).

Dysfunction in the above mentioned areas would result in poor postural control (moderate hypotonia or hypertonia, poor distal control, static and dynamic balance), difficulty in motor learning (learning new skills, planning of movement, adaptation to change, automatization), and poor sensorimotor coordination (coordination within/between limbs, sequencing of movement, use of feedback, timing, anticipation, strategic planning) (Zang et al., 2007).

Balance deficit in children with ADHD/C is either a separate, co-morbid conditions or side effects of dysfunctional attention or impulsiveness. The cooperation of the ADHD children and their ability to attend & understand the task needed represented an important limitation in our study. Geuze (2005) and Fliers et al. (2009) argued a shared etiology for ADHD with co-occurring balance / motor problems that might be attributed to genetic and/or shared environment effects. The postural function has been closely associated not only with gross motor movements, such as sitting, standing, walking and fine motor movements, but also with human behaviors (Shum & Pang, 2009).

## 5. Conclusion

From this work, it is obvious that the static postural control is one of the domains of perceptual motor performance in which a group of children with ADHD/C can be impaired. The studied ADHD/C group was homogenous in terms of severity of symptoms. They showed poor static postural control, especially in extremely difficult situations. The authors assumed that the studied ADHD/C exhibited a form of sensory integration disorder reflected on their postural control.

In light of the current study, it is recommended to follow up the progress of the postural control in the studied children with ADHD/C. History of postural control problems should be included as routine in evaluation of ADHD/C children and referral for postural testing could be done whenever possible. The effects of CNS stimulants in balance improvement in this population warrant to be investigated. Retraining for Balance may be a functional technique for training children and youth with sensorimotor difficulties and might constitute a complement to regular treatment of ADHD, but controlled studies are necessary before more decisive conclusions can be drawn.

## 6. Acknowledgement

The contribution and cooperation of the children parents' that enriched this work was highly appreciated

## 7. References

Allum J. & Shepard N. (1999). An overview of the clinical use of dynamic posturography in the differential diagnosis of balance disorders. *Journal of vestibular research,* vol. 9, pp. 223-52, ISSN 1545-9683.

American Psychiatric Association (1994). *Diagnostic and statistical manual of mental disorders* (4th ed), ISBN-10: 0890420254, Washington DC.

Ayers J. (1972). *Sensory Integration and Learning Disorders.* Western Psychological Services, ISBN-10: 0874243033, Los Angeles.

Bauer S., Woollacott M. & Shumway-Cook A. (2001). The interacting effects of cognitive demand and recovery of postural stability in balance-impaired elderly persons. *The Journal of Gerontology,* vol. 56, No. 8, pp. 489-96, doi: 10.1093/gerona/56.8.M489.

Ben-Sasson A., Carter A. & Briggs-Gowan, M. (2009). Sensory over-responsivity in elementary school: Prevalence and Social-Emotional Correlates. *Journal of Abnormal Child Psychology,* vol. 37, pp.705–16, ISSN-0091-0627

Bledsoe J., Semrud-Clikeman M. & Pliszka S. (2009). A magnetic resonance imaging study of the cerebellar vermis in chronically treated and treatment-naïve children with attention-deficit/hyperactivity disorder combined type. *Biological Psychiatry Journal,* vol.1, No. 65, pp. 620- 4, doi:10.1016/j.biopsych.2008.11.030

Buderath P., Gärtner K., Frings M., Christiansen H., Schoch B., Konczak J. & et al.(2009). Postural and gait performance in children with attention-deficit/hyperactivity disorder. *Gait Posture,* vol. 29, No. 2, pp.249-54.

Bundy A., Murray E. & Lane S. (2002). *Sensory integration theory and practice* (2nd ed), FA Davis Co., ISBN 10: 0803605455, Philadelphia.

Cherng R., Chen J. & Su F. (2001). Vestibular system in performance of standing balance of children and young adults under altered sensory conditions. *Perceptual Motor Skills,* vol. 92, pp. 1167-79, ISSN: 1534-7362

Conner C. (1997). *User manual and administration guide of the Conner's rating scale revised,* multihealth system incorporated.

Dunn W. & Bennett D. (2002). Patterns of sensory processing in children with attention deficit hyperactivity disorder. *Occupational Therapy Journal of Research, vol.* 22, pp. 4-15, ISSN-0276-1599.

Fliers E., Vermeulen S. & Rijsdijk F. (2009). ADHD and poor motor performance from a family genetic perspective. *Journal of the American Academy of Child and Adolescent Psychiatry,* vol. 48, pp. 25–34, ISSN: 0890-8567.

Geuze R. (2005). Postural Control in Children with Developmental Coordination Disorder. *Neural plasticity,* vol.12, pp.183-96, doi:10.1155/NP.2005.183.

Guskiewicz K. & Perrin D. (1996). Research and clinical applications of assessing balance. *Journal of Sport Rehabilitation,* vol. 5, pp. 45-53.

Horak F. (1987). Clinical measurement of postural control in adults. *Physical Therapy*, vol. 67, No. 12, pp.1881-1885.

Hunter M. & Hoffman M. (2001). Postural control: Visual and cognitive manipulations. *Gait Posture*, vol. 13, pp. 41-48.

Ismail M. & Melika L. (1961). *Arabic Wechsler III Intelligence Scale for Children*. Dar Al-Nahda Al-Arabia, Egypt.

Miller L., Anzalone M., Lane S.,Cermak S.& Osten E. (2007a). Concept evolution in sensory integration: a proposed nosology for diagnosis. *The American Journal of Occupational Therapy*, vol.61, pp.135–140.

Miller L., Nielsen D., Schoen S. & Brett-Green B. (2009). Perspectives on sensory processing disorder: a call for translational research. *Frontiers in Integrative Neuroscience*, vol. 3, pp.1-12. doi: 10.3389/neuro.07.029.2009.

Mulligan S. (1996). An analysis of score patterns of children with attention disorders on the sensory integration and Praxis Tests. *The American Journal of Occupational Therapy*, vol. 50, pp. 647-54, ISSN: 0272-9490

Palmeri R., Ingrsoll C., Stone M. & Krause B. (2002). Centre of pressure parameters used in the assessment of postural control. *Journal of Sports Rehabilitation*, vol. 11, pp.51-66.

Niedermeyer E. & Naidu S.(1997). Degenerative disorders of the central nervous system. *Clinical Neurophysiolology*, vol. 102, pp.299–302.

Sheehan D., Lecrubier Y., Sheehan K., Amorim P., Janavs J., Weiller E. & et al. (1998). The Mini-International Neuro-psychiatric Interview (M.I.N.I): The development and validation of a structured diagnostic psychiatric interview for DSM-IV and ICD-10. *Journal of Clinical Psychiatry* vol. 59, pp.22–33, ISSN: 2161-7325

Shepard N. & Janky K. (2008). Background and technique of computerized dynamic posturography. In: *Balance function assessment and management*, Jacobson G. & Shepard N. (eds.): Plural publishing, ISBN-10: 1597561002, San Diego CA.

Shepard N. & Telian S. 1996. *Practical management of the balance disorder patient*. Singular, ISBN: 9781879105843, San Diego, SA.

Shum S. & Pang M. (2009). Children with attention deficit hyperactivity disorder have impaired balance function: involvement of somatosensory, visual, and vestibular systems. *J Pediatr*, vol. 155, pp. 245-9, doi:10.1016/j.jpeds.2009.02.032

Stray L., Stray T., Iversen S., Ruud A., Ellertsen B. & Tønnessen F. (2009). The Motor Function Neurological Assessment (MFNU) as an indicator of motor function problems in boys with ADHD. *Behavioral and Brain Functions*, Vol. 5, pp.22, ISSN 1744-9081.

Wang J., Wang Y. & Ren Y. (2003). A case-control study on balance function of attention deficit hyperactivity disorder (ADHD) children. *Beijing Da Xue Xue Bao*, vol.18, No. 35, pp. 280-3.

Zang Y., Bomei G., Qian Q. & Wang Y. (2002). Objective measurement of the balance dysfunction in attention deficit hyperactivity disorder in children. *Chinese Journal of Clinical Rehabilitation*, vol. 6, pp.1372-1374.

Zang F., He Y., Zhu Z., Cao J., Sui Q., Liang M. & et al. (2007). Altered baseline brain activity in children with ADHD revealed by resting-state functional MRI. *Brain and Development*, vol. 29, No.2, pp. 83-91, ISSN, 0387-7604.

# 2

# Variability, Noise and Predictability in Motor Response Times: Adaptation or Misadaptation?

Tymothée Poitou and Pierre Pouget
*Université Pierre et Marie Curie, ICM, Unité Mixte de Recherche, CNRS UMR 7225, INSERM UMRS 975, Hôpital Salpêtrière, Paris, France*

## 1. Introduction

The ability to maintain a focus of attention on a selected item is crucial for complex and adapted behaviors. In the wild, a predator must be able to track the appearance of other animals while pursuing a prey when the prey must be able to focus on its surrounding environment in order to avoid unexpected obstacles. To survive both predator and prey must share this ability to dissociate the focus of attention and the orientation of gaze. In the context of a laboratory the study of this ability to sustain attention has often been examined using tasks which require individuals to actively maintain performance speed and accuracy over long testing period (von Voss, 1899; Kraepelin, 1902; Robinson & Bills, 1924; Russo & Vignolo, 1967; Rabbitt, 1969; Rabbitt, 1980; Sanders & Hoogenboom, 1970; Richer and Lepage, 1996). In practice, these tasks require subjects to be engaged in repetitive activities such as simply detecting visual objects presented on computer screens. During these repetitive activities performance of subjects varies and evidence concerning the salience of intra-individual-variability to the study of behavior is becoming a compelling reminder that the prevailing emphasis on one of the seemingly most fundamental concepts in traditional differential psychology represents an oversimplification that can hinder the search for powerful and general lawful relationships (Nesselroade, J.R., et al., 2002).

In the last decades, new methodological approaches have improved research in intra-individual-variability. These methods are accounting for both the deterministic and the stochastic components of psychological processes at the intra-individual level. Finally, tools have been developed to account beyond individual differences in variance and covariance of latent variables given measurement invariance. Today, the concerns are now less about whether variability within an individual should be studied than it is about how to make use of this important source of information to assess psychological processes. Authors such as Sliwinski, Almeida, Smuth, & Stawski (2009); MacDonald, Nyberg, & Bäckman (2009) and many others are writing about how to best use both sources of information together to illustrate, how short term variability over time can differ between people in diagnostically interesting ways. In fact, it may be time to make the case that the amount of variability is less a focus than is the time dynamics of the variability particularly in the studies of ADHD.

Ram and colleagues (2009) make even the case that the time structure of intra-individual-variability needs to be considered along with what they call the net variability. Intra-individual-variability is, in some sense, the intellectual parent of dynamical systems analysis in psychology. Because, ADHD subjects demonstrated significantly more variable performance than controls and because numerous studies supports intra-individual variability as a hallmark feature of ADHD beyond the domain of response inhibition and reinforces, there is a crucial need to fully consider variability in ADHD more broadly.

## 2. Intra-individual variability

Typically, the global measures of speed and accuracy are used to determine an individual's ability to sustain attention. However an important limit of all measures reflecting central tendency is that they only coarsely summarize the full response time (RT) distribution, without capturing potentially useful information on intra-individual RT variability (Carpenter and William, 1995; Larson & Alderton, 1990; Rabbitt, Osman, Moore, & Stollery, 2001). In fact, RT distributions are often asymmetrical: they have a steep slope on the left side which is due to a rather narrow range of very fast responses, and they have an elongated right tail, arising from a substantial amount of more broadly distributed slow responses (Leth-Steensen, Elbaz, & Douglas, 2000; Logan, 1992). Recent studies have shown that subjects with attention-deficit/hyperactivity disorder present larger response variability across a variety of speeded-reaction time tasks, laboratories, and cultures (see for review, Castellanos and Tannock, 2002). This high response variability is informative because it may reflect intrinsic properties that extend far beyond the distributional properties of RTs (Gilden and Hancock 2007; Pouget et al. 2010; Pouget et al. 2011). Because RTs are almost always collected in large blocks of trials, the natural ordering of trials generates historical record and the RT records have been shown to have characteristic structures (Gilden, 1997, 2001; Thornton & Gilden, 2005; Van Orden, Holden, & Turvey, 2003, 2005; see also Nelson et al. 2010; Emeric et al. 2007). In particular, it is now well established that RT sequences in normal adults often show evidence of a long-term memory process known as 1/f noise (Gilden, 2001; Thornton & Gilden, 2005), so named because its power spectrum falls inversely with frequency. This kind of noise is found in that part of the data generally regarded as unexplained variance, the trial-to-trial residual variability. More generally, this power-law scaling relation implies that results of a measurement depend on the measurement scale or sampling unit used to take the measurement (over a finite range of scales). Power-law scaling relations, linear relations between the logarithms of the scale and the logarithms of the measurement result, are commonly observed of natural phenomena described using fractal geometry and are symptomatic of self-similar patterns (Bassingthwaighte et al., 1994). In ADHD, the result of a measurement of a natural fractal is also amplified in proportion to the measurement scale Response time variability measures aspects of executive functioning related to a person's ability to consistently focus and purposefully sustain mental effort. With prolonged time on task, work speed has been observed not only to become slower but also less regular (Gilden, 2001; Thornton & Gilden, 2005). For example, von Voss (1899) observed that with prolonged work on a digit addition task, the frequency of long responses increased whereas there was no change in the fastest responses. The question of what causes the characteristic work speed fluctuations is still unresolved (Weissman, Roberts, Visscher, & Woldorff, 2006; but see Gilden, 2001; Thornton

& Gilden, 2005; Pouget et al. 2011). Previous investigations into the nature of intraindividual RT variability drew the conclusion that occasionally occurring attentional lapses may cause the slower responses (e.g., Bertelson & Joffe, 1963; Bills, 1937; Hockey, 1986; Sanders, 1998). The lapses were believed to be involuntary resting pauses, enforced by the accumulation of fatigue during the task (Bertelson & Joffe, 1963; Sanders & Hoogenboom, 1970). This notion was also supported by studies showing that mental fatigue, as induced by prolonged task performance, primarily affects the upper end of the intraindividual RT distribution (Fiske & Rice, 1955; Welford, 1984). In addition, it has been suggested that occasionally occurring task-irrelevant events are often responsible for some of the response time outliers (Jensen, 1992; Smallwood et al., 2004; Ulrich & Miller, 1994), particularly when it is required to maintain performance over extended time periods (Stuss, Meiran, Guzman, Lafleche, & Willmer, 1996; Stuss, Murphy, Binns, & Alexander, 2003).

Recent findings indicate that under conditions requiring higher degrees of response control, increased variability in ADHD is present throughout the RT distribution, regardless of ADHD subtype, reflecting inefficiency in neural mechanisms critical to engaging a state of preparedness to respond (Hervey et al. 2006 ; Castellanos et al. 2005).Children with ADHD, however, do not only have increased RT variances, they also seem to be slower in their mean response times. In many response time tasks, larger mean response times are accompanied by larger response time variances (e.g., Luce, 1986; Wagenmakers & Brown, 2007; Wagenmakers et al., 2005). Most explanations of this phenomenon involve the proposition of an information accumulation process for which this dependence between mean and variance holds naturally (see Luce, 1986; Ratcliff, 1978, Carpenter and Williams 2005; Pouget et al. 2011). For example, a change in information accumulation efficiency then causes a change in mean response times as well as a change in response time variance (Shadlen and Newsome, 1997; Hanes and Schall, 1995; Pouget et al., 2011). Other factors than accumulation efficiency may also influence information processing. Therefore, it is possible that the increased RT variance is, at least partially, due to the same source that causes the overall slower responses.

## 3. Variability and stationarity

As presented in the preceding paragraphs, the literature supports the view that intra-individual RT variability in sustained attention tasks is an empirical phenomenon distinct from other performance characteristics (Pieters, 1985; Sanders, 1983). It is in fact very compelling that in many RT tasks the observed within-person variability is 20% to 50% of the between-person variability when both are expressed in standard deviation units. In development, an increase with age in intra-individual-variability might be expected if fluctuating levels of performance are an early sign of cognitive decline. It is also possible that, for some variables, higher amounts of intra-individual-variability in elderly persons are positive, rather than negative, outcomes. For example, higher variability might signify greater adaptability, less rigidity, or more creativity. Numerous publications on substantive aspects of the topic (e.g., Butler, Hokanson, & Flynn, 1994; Eizenman, Nesselroade, Featherman, & Rowe, 1997; Hertzog, Dixon, & Hultsch, 1992), treatments of pertinent methodological issues are also appearing with rapidity (e.g., Boker & Nesselroade, 2002; Browne & Nesselroade, 2002; Hamaker, Dolan, & Molenaar, 2003; McArdle, 1982; McArdle

& Hamagami, 2001; Molenaar, 1985; Moskowitz & Hershberger, 2002; Nesselroade & Molenaar, 1999; West & Hepworth, 1991).

Just how important does information on intra-individual variability seem to be in the current state of behavioral inquiry? When intra-individual variability in a given attribute is small, the inter-individual differences in that attribute supply the useful information, from a prediction standpoint; when intra-individual variability is large, however, they may not. Indeed, in the latter case, scores from only one occasion can yield highly misleading inter-individual-differences information. From the perspective of classical theory, short term, intra-individual variability is noise. Opposing such negative sentiments are the more positive findings that short-term intra-individual variability is a valid indicator of substantively important events. But the balance between noise and stationarity is fragile. A source of adaptation in some cases too much variability can also lead to dramatic lost of efficacy. Several pieces of evidence suggest that increased intra-subject variability may be a good candidate as an intermediate endophenotype of ADHD (Castellanos & Tannock, 2002, Castellanos, Sonuga-Barke, Milham, &Tannock, 2006). First, increased variability in responding has been demonstrated to correlate with impulsive responding and self-report of inattention to tasks (Rommelse et al., 2007; Simmonds et al., 2007; Strandburg et al., 1996), suggesting that variability in responding is a contributing factor to expression of diagnostic characteristics of ADHD. Further, several studies have demonstrated that close family members of individuals with ADHD demonstrate increased variability in responding, including, siblings sharing an ADHD diagnosis, discordant dizygotic twins, and siblings who do not meet criteria for diagnosis of ADHD (Bidwell, Willcutt, DeFries, & Pennington, 2007; Rommelse et al., 2007). This pattern of results suggests a genetic mechanism for expression of the phenotype. Analyses characterizing intra-individual variability in ADHD has revealed a pattern of occasional responses with unusually long reaction time, with the majority of responses being comparable to comparison groups (Castellanos et al., 2005; Hervey et al., 2006; Leth-Steensen, King Elbaz, & Douglas, 2000).

## 4. Neurophysiological substrate of intra-individual variability

Aside the genetic approaches, cognitive and neurophysiological studies have revealed that candidate endophenotypes in ADHD include inhibitory-based executive deficits associated with frontal–striatal dysfunction (Nigg et al 2005) delay-related motivational processes linked to limbic–ventral striatal circuits (Sonuga-Barke 2002, 2003); cerebellar-based timing deficits (Toplak et al 2003); and posterior parietal noradrenergic orienting deficits (van Leeuwen et al 1998). Given the likely pathophysiologic heterogeneity of ADHD, all these candidates are not mutually exclusive; they could each be playing substantial roles in different clusters within the ADHD groups of patients. At a molecular level, dysfunctional modulation of select neurotransmitters, including those in the catecholamine and ACh systems, gives rise to increased neural noise that might contribute to increased intra-individual variability in cognitive performance. Alterations in the dopamine system are well documented in populations that exhibit increased behavioral intra-individual variability, including the elderly, ADHD children (Bellgrove et al. 2005), schizophrenics and patients with Parkinson's disease. These findings have been substantiated in computational modeling studies showing that reduced dopamine activity increases neural noise, resulting in less distinct cortical representations manifest as decreases in cognitive performance and increases in behavioral intra-individual variability.

At a whole-brain level, functional activation techniques, such as electroencephalograms (EEGs) and functional magnetic resonance imaging (fMRI), are tempting to link behavioral intra-individual variability to brain function. Functional imaging studies in children with ADHD did observe abnormalities in inferior and medial prefrontal, striatal and temporo-parietal brain regions during tasks of interference inhibition (Vaidya et al., 2005; Konrad et al., 2006; Rubia et al., 2007b, 2009c, 2011a; for review see Rubia, 2010). Adults with ADHD when compared to controls in an a priori region of interest, show less activity in anterior cingulate (Bush et al., 1999), while other studies found reduced activation compared to healthy adults in the right inferior prefrontal cortex during event-related interference inhibition trials, but enhanced right medial frontal activation for a blocked interference inhibition condition (Banich et al., 2009). Inconsistent findings of either increased or decreased frontal, parietal, temporal and cingulate activation in adults with ADHD compared to control subjects were also observed in fMRI studies of other executive functions as such motor response inhibition and working memory (Epstein et al., 2007; Banich et al., 2009; Dibbets et al., 2009; Cubillo et al., 2010). The inconsistencies between findings could be related to the fact that most of the published fMRI studies in adult ADHD have included patients with a stimulant medication history (Bush et al., 1999; Valera et al., 2005, 2010a; Hale et al., 2007; Banich et al., 2009; Dibbets et al., 2009; Wolf et al., 2009; Cubillo and Rubia, 2010). Chronic stimulant medication is an important confound given evidence for long-term effects of stimulant medication on brain structure (Bledsoe et al., 2009; Shaw et al., 2009) and function (Konrad et al., 2007 but see Cubillo et al., 2010). But inconsistency could also be related to averaging methods used to analyze these data, while behavioral studies have revealed a critical role in response variability.

Using structural magnetic resonance imaging (MRI) in ADHD patients, recent studies found reduced volume and cortical thickness in inferior prefrontal cortex (IFC) but also other frontal brain regions, as well as parieto-temporal regions, the basal ganglia, the splenium of the corpus callosum, and the cerebellum (McAlonan et al., 2007; Durston et al., 2004; Semrud-Clikeman et al. 2000; Berquin et al. 1998; Mostofsky et al. 1998). Recent analyses of structural data in childhood ADHD have also shown reductions relative to control subjects in posterior inferior vermis of the cerebellum, the splenium of the corpus callosum, total and right cerebral volumes, right caudate, and various frontal regions (Tian et al., 2006). The other meta-analysis was of whole-brain voxel-based morphometry imaging studies, avoiding the a priori bias of region selection, and identified a significant regional gray matter reduction in ADHD children compared with control subjects in right putamen and globus pallidus (Qui et al. 2009; Qiu et al. 2010). Diffusion tensor imaging studies have furthermore provided evidence for abnormalities at the neural network level, showing abnormalities in multiple white matter tracts in cingulate and fronto-striatal, as well as fronto-parietal, fronto-cerebellar, and parieto-occipital white matter tracts, in children, as well as adults, with ADHD compared with comparison subjects (Konrad et al. 2011; Thomason and Thompson 2011; Konrad et al. 2010; Ashtari et al. 2005). Longitudinal imaging studies have provided evidence that the structural abnormalities in these late-developing fronto-striato-cerebellar and frontoparietal systems are due to a late structural maturation of these regions (Rubia, 2011; Rubia et al. 2009; Yang et al. 2007). Thus, the peak of cortical thickness maturation has been shown to be delayed in children with ADHD compared with healthy peers, including frontal and temporal areas (Shaw et al. 2009; Shaw et al. 2007; Durston et al. 2003). All these regions and connections could be part of a network responsible for the variability and stationarity of behaviors, and particular defects on these networks could results in the observed and pathological expression of ADHD.

Finally and to go back to the first describe genetic approach. A strong genetic contribution to ADHD was evidenced through twin, family and adoption studies, and consider- able efforts have been made to identify genes involved in its etiology (for recent review see Cummins et al. 2011; Finke et al. 2011; Semrud-Clikeman et al. 2011). However, results of candidate gene associations for ADHD yielded largely inconsistent results. Dopamine dysregulation is thought to play a crucial role and the dopamine genotypes of DAT1 and dopamine receptor D4 (DRD4) 7-repeat allele are most commonly associated with the disorder (Johnson et al. 2008). The DRD4-7-7 genotype has been associated with reduced volume and cortical thickness of the right IFC in normal development, which was, furthermore, particularly pronounced in ADHD children with the genotype (Semrud-Clikeman et al. 2011). The DAT1 genotypes have been associated with abnormal caudate volume, as well as activation in patients with ADHD (Tovo-Rodrigues et al. 2011; Szobot et al. 2011; Todd et al. 2005). Antisocial behaviors, including psychopathy, have more commonly been associated with serotonin genotypes. Thus, the short allele of the serotonin transporter has been associated with impulsive and antisocial behavior features in alcohol abuse (Li et al. 2010; Herman et al. 2011, see also for review Nordquist and Oreland, 2010) in adults. In healthy adults, it has been related to a dysmorphology and dysregulation of the ventromedial prefrontal cortex, including anterior cingulate and medial frontal cortex, and the amygdala, as well as the functional connectivity between both structures. Abnormal connectivity between amygdala hyperactivity and orbitofrontal hypo-responsivity in relation to negative emotions has been suggested to underlie impulsive aggression (Rubia et al. 2011). Genetic predisposition, hence, may play a role in the development of the disorder-specific dysregulation of IFC-striatal and ventromedial-limbic neural networks in ADHD and antisocial-aggressive behaviors, respectively.

There are still a great number of methodological questions that remain to be addressed in the field of behavior intra-variability. Moment-by-moment fluctuations characteristic of biological processes are fundamentally dynamic in that their quantity and quality of patterning and periodicity are highly sensitive to contextual factors (Stein and Kleiger 1999). However, in many cases, variability is handled by collapsing across time intervals, yielding a single-point estimate of deviation around the mean (SD) for each subject. Group comparisons of variability are then based on group means of individual SD. Thus, although RT studies in ADHD are nearly too numerous to count, the question of the robustness of the association between ADHD and variability has yet to be addressed quantitatively. Significant factors, such as the context within which the organism is working, the tasks being performed, and the internal physiologic and/or cognitive state are affecting intra-individual variability (Borger and Van der Meere 2000; Leung et al 2000; Sonuga-Barke 2003; Swaab-Barneveld et al 2000). For this reason, an analysis of the dynamic properties of ISV requires an examination of the extent to which it is both modifiable and modified by changes in contextual factors. Indeed, intra-invidividual variability might be distinctive not only in terms of amount or degree and its temporal structure and periodicity but also in terms of its relationship to other factors within the environment, as demonstrated by the frequently documented observation that the performance of children with ADHD is highly context dependent (Corkum and Siegel 1993). To address this issue, one need to study the quantitative and qualitative characteristics of intra-individual variability in diverse physiologic states. More systematic investigation of the nature of intra-individual-variability and change in a wide array of attributes is both compelling and timely. The first aspect on which, one needs to focus the methodological and the relevant evaluation of the representativeness of single-occasion assessment. The second aspect relates to whether there are age differences in moment-to-moment, or day-to-day, intra-individual-variability and, if so, what are their salient features.

## 5. Discussion

Just how important does information on intra-individual-variability seem to be in the current state of behavioral inquiry of ADHD? In the last decades, it has been argued that increased intra-individual-variability in cognitive performance could indeed be a valid indicator' of impending cognitive change in children (Eizenman et al. 1997; Rowe and Kahn, 1987; Castelanos et al., 2002). However even more systematic investigation of the nature of intra- individual variability and change in a wide array of attributes is both compelling and timely.

These examinations will be necessary for at least two reasons. To better understand the interactions between three key notions of stability, variability, and adaptability; but also to assess at a statistical and functional levels the normal and pathological dynamics of a given system and its behavior over time. Indeed, understanding the relationship between these concepts constitutes a key issue in research on complex biological systems in fields like human motor control and performance. Stability can be specified either by the property of a system to resist changes, that is, to exhibit minimal variation while facing changing conditions, or by its ability to recover a state of equilibrium after perturbation. This may be contrasted with a dynamic form of stability, which refers to reproducible and predictable *patterns of changes* in the system's functioning under varying internal or external constraints. While static stability implies that the variables determining the system's state are maintained within a limited range, the dynamic definition allows the stability of some global behavior to be maintained by changing states of the system (Ahn, Tewari, Poon, & Phillips, 2006).

A general assumption is that enhanced variability of a given behavior reflects its reduced stability. Therefore, behavioral stability has often been appropriately inferred from the observation of small variance. However, even though stability and variability (as assessed by basic Gaussian statistics) are obviously two related aspects, the invariant nature of this relationship is arguable. One may intuitively wonder, for instance, which of the following two behaviors should be termed "more stable": the behavior that exhibits the smallest fluctuations or the behavior that is perpetuated *in spite of* maximal variability (Riley & Turvey, 2002). In other words and to go back the title of our chapter one may question the origins of the fragile balance between variability, noise and predictability in control of human behavior.

## 6. References

Ahn AC, Tewari M, Poon CS, Phillips RS. The clinical applications of a systems approach. PLoS Med. 2006 Jul;3(7):e209. Epub 2006 May 23

Ashtari M, Kumra S, Bhaskar SL, Clarke T, Thaden E, Cervellione KL, Rhinewine J, Kane JM, Adesman A, Milanaik R, Maytal J, Diamond A, Szeszko P, Ardekani BA. Attention-deficit/hyperactivity disorder: a preliminary diffusion tensor imaging study. *Biol Psychiatry*. 2005 Mar 1;57(5):448-55.

Bassingthwaighte J.B., L.S. Liebowitch and B.J. West, 1994: *Fractal Physiology*, Oxford University Press, Oxford.

Berquin PC, Giedd JN, Jacobsen LK, Hamburger SD, Krain AL, Rapoport JL, et al. Cerebellum in attention-deficit hyperactivity disorder: a morphometric MRI study. *Neurology*. 1998;50:1087–93.

Bertelson P., Joffe R. Blockings in prolonged serial responding. *Ergonomics*. 1963; 6:109–116.

Bidwell, L. C., Willcutt, E. G., McQueen, M. B., DeFries, J. C., & Pennington, B. F. A family-based association study of DRD4, DAT1, and 5HTT and continuous traits of Attention-Deficit/Hyperactivity Disorder. *Behavior Genetics*, 2011, 41 (1), 165-74.

Bills, A. G. (1937). Facilitation and inhibition in mental work. *Psychological Bulletin, 34*, 286-309.

Boker, S. M., & Nesselroade, J. R. A method for modeling the intrinsic dynamics of intraindividual variability: Recovering the parameters of simulated oscillators in multi-wave data. *Multivariate Behavioral Research*, 2002, 37, 127–160.

Butler, A. C., Hokanson, J. E., & Flynn, H. A. A comparison of self- esteem lability and low trait self-esteem as vulnerability factors for depression. *Journal of Personality and Social Psychology*, 1994, 66, 166–177.

Carpenter, R. H. S., and Williams, M. L. L. Neural computation of log likelihood in the control of saccadic eye movements. *Nature*, 1995, 377 59-62.

Castellanos FX, Sonuga-Barke EJ, Milham MP, Tannock R. *Characterizing cognition in ADHD: beyond executive dysfunction. Trends Cogn Sci*. 2006 Mar;10(3):117-23.

Castellanos FX, Tannock R. Neuroscience of attention-deficit/hyperactivity disorder: the search for endophenotypes. *Nat Rev Neurosci*. 2002 Aug;3(8):617- 28.

Cattell, R. B. (1957). *Personality and motivation: Structure and measurement*. New York: World Book.

Corkum, P. V., & Siegel, L. S. Is the continuous performance task a valuable research tool for use with children with attention- deficit-hyperactivity disorder? Journal of Child Psychology and Psychiatry, and Allied Disciplines, 1993, 34, 1217-1239.

Cummins TD, Hawi Z, Hocking J, Strudwick M, Hester R, Garavan H, Wagner J, Chambers CD, Bellgrove MA. Dopamine transporter genotype predicts behavioural and neural measures of response inhibition. *Mol Psychiatry*. 2011 Aug 30.

Durston S, Hulshoff Pol HE, Schnack HG, Buitelaar JK, Steenhuis MP, Minderaa RB, et al. Magnetic resonance imaging of boys with attention-deficit/hyperactivity disorder and their unaffected siblings. *J Am Acad Child Adolesc Psychiatry* 2004;43:332–40.

Durston S, Tottenham NT, Thomas KM, Davidson MC, Eigsti IM, Yang Y, Ulug AM, Casey BJ. Differential patterns of striatal activation in young children with and without ADHD. *Biol Psychiatry.* 2003 May 15;53(10):871-8.

Eizenman, D. R., Nesselroade, J. R., Featherman, D. L., & Rowe, J. W. (1997). Intra-individual variability in perceived control in an elderly sample: The MacArthur Successful Aging Studies. *Psychology and Aging.* 12, 489–502.

Emeric EE, Brown JW, Boucher L, Carpenter RH, Hanes DP, Harris R, Logan GD, Mashru RN, Paré M, Pouget P, Stuphorn V, Taylor TL, Schall JD. Influence of history on saccade countermanding performance in humans and macaque monkeys. *Vision Res.* 2007 Jan;47(1):35-49.

Finke K, Schwarzkopf W, Müller U, Frodl T, Müller HJ, Schneider WX, Engel RR, Riedel M, Möller HJ, Hennig-Fast K. Disentangling the adult attention-deficit hyperactivity disorder endophenotype: Parametric measurement of attention. J *Abnorm Psychol.* 2011 Aug 22.

Fiske, D. W., & Rice, L. Intra-individual response variability. *Psychological Bulletin.* 1955, 52, 217-250.

Gilden, D.L. & Hancock, H. Response Variability in Attention Deficit Disorders. *Psychological Science,* 2007, 18(9), 796-802.

Gilden, D.L. Cognitive emissions of 1/f noise. *Psychological Review,* 2001, 108, 33–56.

Gilden, D.L. Fluctuations in the time required for elementary decisions. *Psychological Science,* 1997, 8, 296–301.

Gilden, D.L., Thornton, T., & Mallon, M.W. 1/f noise in human cognition. *Science,* 1995, 267, 1837–1839.

Herman AI, Conner TS, Anton RF, Gelernter J, Kranzler HR, Covault J. Variation in the gene encoding the serotonin transporter is associated with a measure of sociopathy in alcoholics. *Addict Biol.* 2011 Jan;16(1):124-32.

Hertzog, C., Dixon, R. A., & Hultsch, D. F. Intraindividual change in text recall of the elderly. *Brain and Language,* 1992, 42, 248–269.

Hervey, A.S., Epstein, J.N., Curry, J.F., Tonev, S., Arnold, L.E., Conners, C.K., et al. Reaction time distribution analysis of neuropsychological performance in an ADHD sample. *Child Neuropsychology,* 2006, *12,* 125–140.

Hockey, G. R. J. Operator efficiency as a function of effects of environmental stress, fatigue, and circadian rhythm. In K. R. Boff, L. Kaufman & J. P. Thomas (Eds.), *Handbook of perception and human performance* (pp. chap. 44). 1986, New York: Wiley.

Jensen, A. R. The importance of intraindividual variability in reaction time. *Personality and Individual Differences,* 1992, 13, 869–882.

Johnson KA, Kelly SP, Robertson IH, Barry E, Mulligan A, Daly M, Lambert D, McDonnell C, Connor TJ, Hawi Z, Gill M, Bellgrove MA. Absence of the 7-repeat variant of the DRD4 VNTR is associated with drifting sustained attention in children with ADHD but not in controls. *Am J Med Genet B Neuropsychiatr Genet.* 2008 Sep 5;147B(6):927-37.

Konrad A, Dielentheis TF, El Masri D, Bayerl M, Fehr C, Gesierich T, Vucurevic G, Stoeter P, Winterer G. Disturbed structural connectivity is related to inattention and impulsivity in adult attention deficit hyperactivity disorder. *Eur J Neurosci.* 2010 Mar;31(5):912-9.

Konrad A, Dielentheis TF, Masri DE, Dellani PR, Stoeter P, Vucurevic G, Winterer G. White matter abnormalities and their impact on attentional performance in adult attention-deficit/hyperactivity disorder. *Eur Arch Psychiatry Clin Neurosci.* 2011 Aug 31.

Kraepelin E. Clinical Psychiaty (translated by Defendorf R.) 1992, London : Macmillan.

Kuntsi, J. and Stevenson, J. Psychological mechanisms in hyperactivity: II. The role of genetic factors. *J. Child Psychol. Psychiatry.* 2001 42, 211–219.

Larson, G.E., & Alderton, D.L. Reaction time variability and intelligence : A « worst performance » analysis of individual differences. *Intelligence*, 1990, 14, 309-325.

Leth-Steensen, C., King Elbaz, Z., & Douglas, V. I. Mean response time, variability, and skew in the responding of ADHD children: A response time distributional approach. *Acta Psychologica*, 2000 104, 167-190.

Li JJ, Lee SS. Latent class analysis of antisocial behavior: interaction of serotonin transporter genotype and maltreatment. *J Abnorm Child Psychol.* 2010 Aug;38(6):789-801.

Logan, G. D. (1992). Shapes of reaction time distributions and shapes of learning curves: A test of the instance theory of automaticity. *Journal of Experimental Psychology: Learning, Memory and Cognition, 18*, 883-914.

Luce, R. D. *Response times*. New York: 1986, Oxford University Press.

MacDonald, S.W.S., Nyberg, L., & Bäckman, L. Intraindividual variability in behavior: Links to brain structure, neurotransmission, and neuronal activity. *Trends in Neurosciences*, 2006, 29, 474-480.

McAlonan GM, Cheung V, Cheung C, Chua SE, Murphy DG, Suckling J, Tai KS, Yip LK, Leung P, Ho TP. Mapping brain structure in attention deficit-hyperactivity disorder: a voxel-based MRI study of regional grey and white matter volume. *Psychiatry Res.* 2007, Feb 28;154(2):171-80.

Molenaar, P. C. M. A dynamic factor model for the analysis of multivariate time series. *Psychometrika*, 1985, 50, 181–202.

Moskowitz, D. S., & Hershberger, S. L. (Eds.). Modeling intraindividual variability with repeated measures data. Mahwah, 2002, NJ: Erlbaum.

Mostofsky SH, Reiss AL, Lockhart P, Denckla MB. Evaluation of cerebellar size in attention-deficit hyperactivity disorder. *J Child Neurol* 1998;13:434–9.

Nelson MJ, Boucher L, Logan GD, Palmeri TJ, Schall JD. Nonindependent and nonstationary response times in stopping and stepping saccade tasks. *Atten Percept Psychophys.* 2010 Oct;72(7):1913-29.

Nesselroade, JR, McArdle, JJ, Aggen, SH, & Meyers, J. Alternative Dynamic Factor Models for Multivariate Time-Series Analyses. In D. M. Moscowitz & S. L. Hersh- berger (Ed.), Modeling intraindividual variability with repeated measures data: Advances and techniques (pp. 235–265). Mahwah, NJ: 2002, Lawrence Erlbaum Associates.

Nigg JT, Willcutt EG, Doyle AE, Sonuga-Barke EJS. Causal heterogene- ity in ADHD: Do we need neuropscychologically impaired subtypes? Biol Psychiatry 2005, 57:1224-1230.

Nordquist N, Oreland L. Serotonin, genetic variability, behaviour, and psychiatric disorders. *Ups J Med Sci*. 2010 Feb;115(1):2-10.

Pieters J. P. M. Reaction time analysis of simple mental tasks: A general approach. *Acta Psychologica*. 1985;59:227–269.

Pouget P, Wattiez N, Rivaud-Péchoux S, Gaymard B. A fragile balance: perturbation of GABA mediated circuit in prefrontal cortex generates high intraindividual performance variability. *PLoS One*. 2009;4(4):e5208.

Qiu A, Crocetti D, Adler M, Mahone EM, Denckla MB, Miller MI, Mostofsky SH. Basal ganglia volume and shape in children with attention deficit hyperactivity disorder. *Am J Psychiatry*. 2009 Jan;166(1):74-82.

Qiu MG, Ye Z, Li QY, Liu GJ, Xie B, Wang J. Changes of Brain Structure and Function in ADHD Children. *Brain Topogr*. 2010 Dec 30.

Rabbitt P. Psychological refractory delay and response-stimulus interval duration in serial, choice-response tasks. *Acta Psychol* 1969; 30: 195-219.

Rabbitt P. The effects of R-S interval duration on serial choice reaction time: preparation time or response monitoring time? *Ergonomics* 1980; 23: 65-77.

Rabbitt, P., Osman, P., Moore B., and Stollery, R. There are stable individual differences in performance variability, both from moment to moment and from day to day. *Quarterly Journal of Experimental Psychology: Human Experimental Psychology* 2001, pp. 981–1003.

Ram, N., & Gerstorf, D. Time-structured and net intraindividual variability: Tools for examining the development of dynamic characteristics and processes. *Psychology and Aging*, 2009, 24, 778-791.

Ratcliff R 1978 A theory of memory retrieval. *Psychological Review* 85: 59–108

Richer F, Lepage M. Frontal lesions increase post-target interference in rapid stimulus streams. *Neuropsychologia* 1996.

Riley MA, Turvey MT. Variability of determinism in motor behavior. *J Mot Behav*. 2002 Jun;34(2):99-125.

Robinson, E. S., & Bills, A. G. Two factors in the work decrement. *Journal of Experimental Psychology*, 1924, 9, 415-443.

Rommelse, N. N., Altink, M. E., Oosterlaan, J., Buschgens, C. J., Buitelaar, J., de Sonneville, L. M., et al. Motor control in children with ADHD and non-affected siblings: Deficits most pronounced using the left hand. *Journal of Child Psychology and Psychiatry*, 2007, 48, 1071- 1079.

Rowe, J. W., & Kahn, R. L. Human aging: Usual and successful. *Science*, 1987, 237, 143–149.

Rubia K, Halari R, Cubillo A, Mohammad AM, Brammer M, Taylor E. Methylphenidate normalises activation and functional connectivity deficits in attention and motivation networks in medication-naïve children with ADHD during a rewarded continuous performance task. *Neuropharmacology*. 2009 Dec;57(7-8):640-52. Epub 2009 Aug 26.

Rubia K. "Cool" inferior frontostriatal dysfunction in attention-deficit/hyperactivity disorder versus "hot" ventromedial orbitofrontal-limbic dysfunction in conduct disorder. *Biol Psychiatry*. 2011 Jun 15;69(12):e69-87.

Russo, M., & Vignolo, A. L. Visual figure-ground discrimination in patients with unilateral cerebral disease. *Cortex*, 1967, 3, 113-127.

Sanders, A. F. and W. Hoogenboom. On the effects of continuous active work on performance, *Acta Psychologica* 1970, 33, 414431.

Sanders, A. F. *Elements of human performance*. Mahwah, NJ: 1998, Lawrence Erlbaum.

Sanders, A. F. Towards a model of stress and human performance. Acta *Psychologica*, 1983, 53, 61-97.

Semrud-Clikeman M, Bledsoe J. Updates on Attention-Deficit/Hyperactivity Disorder and Learning Disorders. *Curr Psychiatry Rep*. 2011 Jun 24.

Semrud-Clikeman M, Steingard RJ, Filipek P, Biederman J, Bekken K, Renshaw PF. Using MRI to examine brain-behavior relationships in males with attention deficit disorder with hyper- activity. *J Am Acad Child Adolesc Psychiatry*. 2000;39:477–84.

Shaw P, Eckstrand K, Sharp W, Blumenthal J, Lerch JP, Greenstein D, Clasen L, Evans A, Giedd J, Rapoport JL. Attention-deficit/hyperactivity disorder is characterized by a delay in cortical maturation. *Proc Natl Acad Sci U S A*. 2007 Dec 4;104(49):19649-54. Epub 2007 Nov 16.

Shaw P, Rabin C. New insights into attention-deficit/hyperactivity disorder using structural neuroimaging. Curr Psychiatry Rep. 2009 Oct;11(5):393-8.

Siegler, R. S. Cognitive variability: A key to understanding cognitive development. *Current Directions in Psychological Science*, 1994, 3, 1–5.

Simmonds DJ, Fotedar SG, Suskauer SJ, Pekar JJ, Denckla MB, Mostofsky SH. Functional brain correlates of response time variability in children. *Neuropsychologia*. 2007;45:2147–2157.

Sliwinski, M. J., Almeida, D. M., Smuth, J., & Stawski, R. S. Intraindividual change and variability in daily stress processes: Findings from two diary burst studies. *Psychology & Aging*, 2009, 24(4), 828-840.

Smallwood, J., Davies, J. B., Heim, D., Finnigan, F., Sudberry, M., O'Conner, R. Subjective experience and the attentional lapse: Task engagement and disengagement during sustained attention. *Consciousness and Cognition*. 2004, *13*, 657-690.

Sonuga-Barke EJS, Saxton T, Hall M. The role of interval underestima- tion in hyperactive children's failure to suppress responses over time. *Behav Brain Res*. 1998, 94:45–50.

Sonuga-Barke EJS. Psychological heterogeneity in AD/HD—a dual pathway model of behaviour and cognition. *Behav Brain Res*. 2002, 130:29 -36.

Stein and Kleiger, 1999 P.K. Stein and R.E. Kleiger, Insights from the study of heart rate variability, *Annu. Rev. Med*. 50 (1999), pp. 249–261.

Strandburg RJ, Marsh JT, Brown WS, Asarnow RF, Higa J, Harper R, Guthrie D. Continuous-processing-related event-related potentials in children with attention deficit hyperactivity disorder. *Biol Psychiatry*. 1996, 15;40(10): 964-80

Stuss DT, Meiran N, Guzman A, Lafleche G, Wilmer J. Do long tests yield a more accurate diagnosis of dementia than short tests ? *Arch Neurol*. 1996 ; 53 : 1033-1039.

Stuss DT, Murphy KJ, Binns MA, Alexander MP. Staying on the job: the frontal lobes control individual performance variability. *Brain*. 2003 Nov; 126(Pt 11):2363-80.

Szobot CM, Roman T, Hutz MH, Genro JP, Shih MC, Hoexter MQ, Júnior N, Pechansky F, Bressan RA, Rohde LA. Molecular imaging genetics of methylphenidate response in ADHD and substance use comorbidity. *Synapse*. 2011 Feb;65(2):154-9.

Thomason ME, Thompson PM. Diffusion imaging, white matter, and psychopathology. *Annu Rev Clin Psychol*. 2011 Apr;7:63-85.

Thornton, T.L., & Gilden, D.L. Provenance of correlations in psychological data. *Psychonomic Bulletin & Review*. 2005, 12, 409–441.

Thouless, R. H. Test unreliability and function fluctuation. *British Journal of Psychology*. 1936, 26, 325–343.

Tian LX, Jiang TZ, Wang YF, Zang YF, He Y, Liang M, et al. Altered resting-state functional connectivity patterns of anterior cingulate cortex in adolescents with attention deficit hyperactivity disorder. *Neurosci Lett*. 2006;400:39–43.

Todd RD, Huang H, Smalley SL, Nelson SF, Willcutt EG, Pennington BF, Smith SD, Faraone SV, Neuman RJ. Collaborative analysis of DRD4 and DAT genotypes in population-defined ADHD subtypes. *J Child Psychol Psychiatry*. 2005 Oct;46(10): 1067-73.

Toplak ME, Rucklidge JJ, Hetherington R, John SCF, Tannock R. Time perception deficits in attention-deficit/hyperactivity disorder and co- morbid reading difficulties in child and adolescent samples. *J Child Psy- chol Psychiatry*. 2003, 44:888 -903.

Tovo-Rodrigues L, Rohde LA, Roman T, Schmitz M, Polanczyk G, Zeni C, Marques FZ, Contini V, Grevet EH, Belmonte-de-Abreu P, Bau CH, Hutz MH. Is there a role for rare variants in DRD4 gene in the susceptibility for ADHD? Searching for an effect of allelic heterogeneity. *Mol Psychiatry*. 2011, Mar 15.

Ulrich, R., & Miller, J. Effects of truncation on reaction time analysis. *Journal of Experimental Psychology: General*, 1994, 123, 34-80.

van Leeuwen TH, Steinhausen HC, Overtoom CC, Pascual-Marqui RD, van't Klooster B, Rothenberger A, et al (1998): The continuous performance test revisited with neuroelectric mapping: Impaired orienting in children with attention deficits. *Behav Brain Res*. 1998, 94:97–110.

van Orden, G.C., Holden, J.G., & Turvey, M.T. Human cognition and 1/f scaling. *Journal of Experimental Psychology: General*, 2005, 134, 117–123.

van Orden, G.C., Holden, J.G., & Turvey, M.T. Self-organiza- tion of cognitive performance. *Journal of Experimental Psychology: General*, 2003, 132, 331–350.

von Voss, G. (1899). Über Schwankungen der geistigen Arbeitsleistung [On the variance of mental performance]. *Psychologische Arbeiten. 1899, 2*, 399-449.

Wagenmakers, E.-J., & Brown, S. On the linear relation between the mean and the standard deviation of a response time distribution. *Psychological Review*, 2007, 114, 830-841.

Wagenmakers, E.-J., Grasman, R., & Molenaar, P. C. M. On the relation between the mean and the variance of a diffusion model response time distribution. *Journal of Mathematical Psychology*, 2005, 49, 195-204.

Weissman, D.H., Roberts, K.C., Visscher, K.M. & Woldorff, M.G. The neural bases of momentary lapses in attention. *Nature Neuroscience*, 2006, 9, 971-978.

Welford, A.T. Between bodily changes and performance: some possible reasons for slowing with age. *Exp Aging Res*. 1984, 10, 73-88.

Wellington TM, Semrud-Clikeman M, Gregory AL, Murphy JM, Lancaster JL. Magnetic resonance imaging volumetric analysis of the putamen in children with ADHD: combined type versus control. *J Atten Disord*. 2006 Nov;10(2):171-80.

West, S. G., & Hepworth, J. T. (1991). Statistical issues in the study of temporal data: Daily experiences. *Journal of Personality*. 1991, 9, 609–662.

Westerberg, H. Visuo-spatial working memory span: a sensitive measure of cognitive deficits in children with ADHD. *Child Neuropsychol.* 2004, 10, 155–161

Woodrow, H. Quotidian variability. *Psychological Review.* 1932, 39, 245–256.

Yang B, Chan RC, Zou X, Jing J, Mai J, Li J. Time perception deficit in children with ADHD. *Brain Res.* 2007 Sep 19;1170:90-6.

3

# Endocrine Dysfunction and Growth in Attention Deficit Hyperactivity Disorder

J. Paul Frindik
*University of Arkansas for Medical Sciences, Little Rock, Arkansas USA*

## 1. Introduction

Scientific literature contains numerous accounts of possible hypothalamic - pituitary axis dysfunction in pediatric patients with various behavioral disorders and/or psychopathology. However, demonstrating any consistent pattern of endocrine dysfunction in these behavioral areas, if indeed such even exists, remains elusive. This chapter will review the literature regarding possible endocrine dysfunction in pediatric patients with ADHD and will focus on the two areas where publications have been the most prolific regarding possible interactions of hormones and behavior: (1) the hypothalamic - pituitary - adrenocortical axis and ADHD and (2) growth in children with Attention Deficit Hyperactivity Disorder. The chapter will conclude with a review of growth hormone therapy in children with ADHD, focusing primarily on ADHD children referred for poor growth and growth evaluation to a pediatric endocrine practice.

## 2. Hypothalamic - Pituitary - Adrenocortical Axis (HPAA) in behavior and ADHD

Variations in hypothalamic - pituitary - adrenocortical axis (HPAA) activity have been associated with or implied as a potential causative factor in a variety of psychopathology. Higher cortisol levels may be seen with aggressive/impulsive behavior whereas lower cortisol levels correlate more often with callous unemotional (CU) behavior. For example, significantly higher adrenal activity, both baseline and stress challenged, is found in five year old boys exhibiting hyperactivity, impulsivity and emotional difficulties than is found in boys without such behavior (Hatzinger et al., 2007). In reviewing older children and adolescents, Barzman et al also found higher cortisol levels to occur more often in conjunction with impulsive aggressive behavior, and lower cortisol levels with CU traits (Barzman et al., 2010). Similarly, Stadler et al. found lower stressed salivary cortisol responses in ADHD children who exhibited CU traits compared to ADHD children with lower CU traits (Stadler et al., 2011).

### 2.1 Dysregulation of Hypothalamic - Pituitary - Adrenocortical Axis Activity in behavior and ADHD

Hypothalamic - Pituitary - Adrenocortical Axis (HPAA) activity is known to vary with gender and age; gender specific differences in HPAA activity are found in children as young as five years old, with girls having higher baseline and stressed adrenal activity compared to boys (Hatzinger et al., 2007). Recent studies increasingly support the idea that at least some varieties of ADHD are associated with dysregulation of the HPAA axis, particularly a reduced response to stressful stimuli (Ma et al., 2011, Anu-Katriina et al., 2011). Interestingly, similar reduced HPAA activity may be also found in people following significant acute traumatic events or with chronic stress (Anu-Katriina et al., 2011).

In some studies, ADHD subtypes are found to have both decreased baseline, wakening cortisol levels and a decreased response to stressful stimuli (Freitag et al., 2009). Freitag et. al. found these decreased wakening cortisol levels in ADHD children only when ADHD was also associated with comorbid oppositional defiant disorder (ODD). Diminished wakening cortisol was not observed in children with ADHD only or in children with ADHD plus conduct disorder or anxiety disorder (Freitag et al., 2009). In other studies, in this case 170 elementary school-age ADHD males, no differences in baseline, waking cortisol levels between ADHD subtypes could be demonstrated (Hastings et al., 2009).

In another study of 128 ADHD male children, age 6 – 14 years, Ma et. al. found that ADHD children with hyperactive impulsive traits (ADHD-HI) had significantly lower baseline (8:00 am) cortisol levels than in children with either ADHD-predominantly inattention type (ADHD-I) or ADHD-combined type (ADHD-C) (Ma et al., 2011). Further, the ADHD group as a whole had significantly lower baseline (8:00 am) cortisol levels compared to non-ADHD controls (226.47±129.12 nmol/L ADHD vs. 384.53±141.43 nmol/L controls, $P<0.001$. Despite these differences in cortisol levels, there were no significant corresponding differences in ACTH levels between the ADHD group as a whole vs. the non-ADHD control group ($P>0.05$).

Cortisol, a hormonal product of the adrenal glands, can be used as an indicator of hypothalamic - pituitary - adrenal -axis (HPAA) activity. In the above referenced study, lower cortisol levels in the ADHD group compared to non-ADHD controls without concurrent differences in ACTH secretion suggest a general disinhibition or dysfunction (under-reactivity) of HPAA activity in ADHD children as opposed to the general population. Finally, no differences were found in ACTH values between the various sub-groups of ADHD ($P>0.05$), despite the lower cortisol values found in the ADHD-HI subgroup, suggesting that even among patients with ADHD, some exhibit greater HPAA disinhibition than others (Ma et al., 2011).

Mothers of 272 eight-year-old children with ADHD were asked to rate their child's ADHD related behavior using such standard methods as the ADHD-IV Rating Scale and the Child Behavior Checklist (CBCL). The mothers' responses were then compared to diurnal salivary cortisol levels from the children in an attempt to correlate behavior with baseline, unstressed salivary cortisol. In contrast to some of the previously referenced studies, no correlations nor significant associations were found between unstressed diurnal cortisol levels and behavioral symptoms. Stressed, salivary cortisol responses to the Trier Social Stress Test for Children (TSST-C) were next evaluated, and differential stress responses were found, again suggesting dysregulated HPAA activity. Lower cortisol stress responses were seen in ADHD males with higher degrees of inattentive symptoms (ADHD-I) than in other ADHD males. Similarly, ADHD-I females had a more rapid fall in stress cortisol response than did other non-ADHD-I females (Anu-Katriina et al., 2011).

Perhaps not surprising and again emphasizing mind-body connection. increased stressed cortisol levels in response to venipuncture are greater in some ADHD children with coexisting anxiety disorders than in those with comorbid disruptive behavior or oppositional problems (Hastings et al., 2009).

A summary of selected studies of the hypothalamic - pituitary - adrenocortical axis activity and associated behavior is presented in Table 1. The table compares the patient populations studied, the endocrine activity measured, associated diagnosis or behavioral traits selected by the researchers, and the qualitative results. Qualitative results are presented, rather than quantitative findings, in order to better compare and appreciate the differences in the studies.

### 2.2 Summary and therapeutic considerations

In summary, studies of hypothalamic - pituitary - adrenocortical axis activity in ADHD children have yielded conflicting results, likely due to the confounding variables of comorbidities, subtypes and heterogeneity of ADHD (Hastings et al., 2009); the variety of methods used to assess stress hormone activity; and current psychosocial conflicts such as parenting issues, family conflicts and stressful, acute life events, all of which can increase cortisol levels (Freitag et al., 2009) and affect the outcome of studies of hormonal levels and behavior. If further research is able to better clarify this complex interaction, measurements of hormonal levels may become a useful diagnostic adjunct to help select the most appropriate treatment modalities for pediatric behavioral disorders and aggression (Barzman et al., 2010).

| Reference | Patient Population | Endocrine | Diagnosis or Associated Behavioral Traits | Result |
|---|---|---|---|---|
| Hatzinger et al., 2007 | 102 five-year olds, (59 boys, 43 girls) | Baseline adrenal activity | hyperactivity, impulsivity and emotional problems (boys); positive emotions (girls) | Increased adrenal activity vs. controls |
| | 102 five-year olds, (59 boys, 43 girls) | Stress challenged adrenal activity | hyperactivity, impulsivity and emotional problems (boys); positive emotions (girls) | Increased adrenal activity vs. controls |
| Freitag et al., 2009 | 128 ADHD age 6–13 years; 96 control age 6–12 years | Cortisol awakening response | ADHD with and without comorbid oppositional defiant (ODD), conduct, or anxiety disorders | Decreased cortisol in ADHD with ODD group vs all other subgroups or controls |
| | 128 ADHD age 6–13 years | Cortisol awakening response | ADHD with and without current psychosocial risk / stress factors | Increased cortisol with adverse parenting, family conflicts, acute life events |
| Hastings et al., 2009 | 170 ADHD boys, elementary school-age | Salivary cortisol at waking, baseline | ADHD with and without comorbid disruptive behavior and anxiety disorders | No differences in wakening, baseline cortisol levels between subgroups |

| | | | | |
|---|---|---|---|---|
| | 170 ADHD boys, elementary school-age | Salivary cortisol post stress | ADHD with and without comorbid disruptive behavior and anxiety disorders | Increased post stress cortisol in ADHD with anxiety disorders |
| Barzman et al., 2010 | children and adolescents | Cortisol | impulsive, aggressive traits | Increased cortisol concentrations |
| | children and adolescents | Cortisol | callous, unemotional traits | Decreased cortisol concentrations |
| Stadler et al., 2011 | 36 boys with ADHD, age 8- 14 years | Salivary cortisol, pre and post stress | higher callous, unemotional traits vs. lower CU traits | Decreased HPAA activity in higher ADHD-CU vs. lower ADHD-CU |
| Ma et al., 2011 | boys with ADHD, 6–14 years old | Baseline 8:00 am Cortisol | ADHD with hyperactive (HI); inattention (I); or combined traits | Lowest baseline cortisol in ADHD-HI vs. ADHD-I or ADHD combined traits |
| | boys with ADHD, 6–14 years old | ACTH levels | ADHD with hyperactive (HI); inattention (I); or combined traits | No difference in ACTH between subgroups |
| | boys age 6–14 years old | Baseline 8:00 am Cortisol | ADHD vs. non-ADHD controls | Lower baseline cortisol in ADHD vs. non-ADHD |
| | boys age 6–14 years old | ACTH levels | ADHD vs. non-ADHD controls | No difference in ACTH in ADHD vs. controls |
| Anu-Katriina et al., 2011 | 272 ADHD, age eight years | Diurnal salivary cortisol | ADHD-IV Rating Scale and the Child Behavior Checklist (CBCL) | No associations between behavioral symptoms and diurnal cortisol levels. |
| | 272 ADHD, age eight years | Salivary cortisol post stress | ADHD-IV Rating Scale, CBCL, Trier Social Stress Test for Children (TSST-C) | Lower stressed cortisol in boys and girls with predominant ADHD-I |

Abbreviations: ADHD = attention deficit hyperactivity disorder; ADHD-CU = ADHD with callous unemotional traits; ADHD-HI = ADHD with hyperactive impulsive traits; ADHD-I = ADHD-predominantly inattention type; ADHD-C = ADHD-combined type

Table 1. Hypothalamic - Pituitary - Adrenocortical Axis (HPAA) Activity in Behavior and ADHD

## 3. Growth in children with ADHD

### 3.1 Growth in non-medically treated ADHD children

Children with ADHD, with or without stimulant therapy, may have significantly decreased height, head circumference, percent body fat, and abdominal circumference,(Ptacek et al., 2009a, 2009b). In two studies Ptacek and colleagues demonstrated decreased nutritional status and lower stature in ADHD children compared to population normals. First, a comparison of non-medicated ADHD boys (n=46, average age 11.03 years) found significantly decreased percent body fat and height in a non-treated ADHD population (Ptacek et al., 2009a). Next, anthropometric measurements, including height, head circumference, percent body fat, and abdominal circumference, were determined in both medically-treated and non-treated ADHD boys (n=104, age 4-16 years). These measurements in the treated and non-treated groups were compared to a normal, non-ADHD population and demonstrated decreased percent body fat and abdominal circumference (markers for nutritional status) as well as decreased height and head circumference in the ADHD population vs. the population norms ($p<0.01$) regardless of treatment status (Ptacek et al., 2009b), suggesting that dysregulated growth may be caused by or at least associated with ADHD itself, independent of other factors.

In a review of 124 ADHD boys, height deficits were found only in early adolescent aged boys with ADHD. This early adolescent decline in growth then spontaneously improved by later adolescence or adulthood with no apparent effect on adult height. Although 89% of the ADHD patients had a history of pharmacologic therapy at some point, height deficits were unrelated to stimulant medication use or weight loss. The mean height SDS of the ADHD group was 0.21 compared to 0.47 in the controls ($p = 0.03$), and the adolescent aged related decline amounted to a mean height deficit of 2.1 cm compared to controls (Spencer et al., 1998).

### 3.2 Growth in stimulant medication treated ADHD children

Reductions in height, weight and / or body mass index are commonly reported with stimulant therapy use for ADHD (Faraone et al., 2008, Negrao & Viljoen, 2011), and may be worse in larger children, those who are naive to therapy or who have a greater cumulative exposure to stimulants (Faraone et al., 2010). Ptacek et al. found a lower percent body fat in medically treated ADHD males (age 4 - 16 years) as compared to non-treated peers (Ptacek et al., 2009b).

A quantitative review of published, longitudinal studies of stimulant-treated ADHD children found significant delays in height and weight over at least of year of treatment time (Faraone et al., 2008). To determine if such decreased growth also occurs in typical clinical scenarios in which patients are treated for longer periods of time in outpatient settings, the heights and weights of 84 ADHD children treated for two - three years continuously with methylphenidate from two community pediatric practices were examined. Growth parameters in treated ADHD children were compared to those of near-same age-matched, healthy siblings who did not have ADHD nor any other chronic condition. Height standard deviation (SD) scores were determined from at least one year prior to starting therapy and thereafter during therapy. Heights were similar pretreatment between boys with ADHD and their age-matched, non-ADHD siblings. During treatment, the height SD scores and growth velocity of the methylphenidate treated boys declined while those of their siblings did not decline and remained constant or even increased. Methylphenidate treated girls showed similar decline in growth. Over time, treated ADHD boys lost about 0.5 SD in height (roughly 3-4 cm), and treated girls lost about 0.6 SD (3-4 cm) in height compared to age-, time- and gender-matched siblings. A clear dose-response curve of methylphenidate dose vs. growth decline was not found, and the authors did not address growth in non-stimulant medication treated ADHD children. However, this study did document a clear decline in growth over time in treated children seen in general practice (Lisska & Rivkees, 2003).

Other studies have found more mild and transient growth disruptions in ADHD children (Negrao & Viljoen 2011), and that even with stimulant treatment, associated growth deceleration and loss of expected height may amount to only a one to two centimeter deficit per year during treatment (Poulton, 2005, Drappatz et al., 2006). Comparison of paired height measurements, before and after ADHD treatment, suggest a height deficit of only about 1 cm/year during the first 1 to 3 years of ADHD treatment (Pliszka, 1998). Growth delay may also be dose dependent during treatment, normalizing afterwards (Faraone et al., 2008) with no long-term adverse effect even with high dose, aggressive stimulant therapy (Pliszka, 1998).

### 3.3 Side effects of stimulant medications and possible causes of growth delay

One of the most frequently reported side effects of stimulant medication use for ADHD is loss of appetite. Other commonly reported and usually relatively minor complaints include headaches, disturbances in mood and other emotional problems, stomach upset, sleep disturbances, and rashes (Tobaiqy et al., 2011). Far less common, but obviously more serious, concerns regarding stimulant medication use include suicidal ideation and sudden cardiac death (Graham et al., 2011).

Interestingly, although a controversial area among physicians and the subject of multiple scientific studies, growth delay was not a reported concern in a survey of parents of children and adolescents receiving stimulant medical therapy (methylphenidate) for ADHD (Tobaiqy et al., 2011).

The exact mechanism by which stimulant medication affects growth is unclear, but possibly includes decreased appetite with subsequent poor weight gain (Poulton, 2005). Since decreased or loss of appetite is a common complaint, It is tempting to ascribe decreased stature as a secondary consequence of stimulant-induced malnutrition. However, despite loss of appetite being reported in as many as 34.3% of treated ADHD patients (Tobaiqy et al., 2011), actual weight loss is not a consistent finding among ADHD studies. While varying degrees of weight loss are found in some patient populations, other studies report no weight loss. Similarly, it has not been possible to convincingly demonstrate an association between decreased nutrition and subsequent poor height growth in ADHD reports (Spencer et al., 1998).

Other possible mechanisms for poor growth include decreased bone mineralization, maturation and linear growth (Lahat et al., 2000), or a medication-induced disruption of the hypothalamic - pituitary - IGF-I axis (Negrao & Viljoen, 2011). Adrenergic stimulants are known to increase dopamine and noradrenaline in neural synapses. Increased dopamine and noradrenaline may in turn inhibit the secretion of growth hormone (GH) and other growth-related peptides including thyroid hormones, sex hormones, prolactin and insulin, ultimately resulting in growth suppression (Negrao & Viljoen, 2011). However, studies of growth hormone have not shown a consistent altered pattern of GH secretion in association with stimulant treatment (Spencer et al.,1998). Finally, dysregulation of various neurotransmitters, in particular the catecholamines, is associated with ADHD itself. Such dysregulation may alter hypothalamic - pituitary function and cause growth delay, even in the absence of stimulant treatment (Spencer et al.,1998).

### 3.4 Concerns regarding growth studies in ADHD

Despite many, and at times, conflicting reports of slow growth in children and adolescents with ADHD (Frindik et al., 2009) and theoretical concerns of adverse interactions between ADHD and the hypothalamic-pituitary-IGF-I axis (Jensen & Garfinkel, 1988), no intrinsic defect or consistent dysfunction in this axis in ADHD has been identified. However, the following caveats regarding studies of growth in ADHD must be kept in mind. Rigorous scientific investigations of growth in children with ADHD have been relatively rare, the majority of ADHD studies are of a retrospective nature and lack significant power, adequate controls, sufficient follow up time, or stringent statistical analysis to draw firm conclusions (Poulton, 2005, Drappatz et al., 2006). Identifying appropriate age, gender, socio-economic, and diagnosis-matched control populations to study growth in ADHD patients with and without stimulant treatment may be especially problematic due to inherent treatment bias. The use of stimulant medication in children with ADHD varies by gender, race, age, language spoken in the home, insurance status, and contact with health care providers (Visser et al., 2007). The frequency with which ADHD medication is used for any reason also varies by geographic location, ranging from 2.1% of the general pediatric population in California to 6.5% in Arkansas (CDC, 2003). ADHD medical treatment is more likely to be used in males than females of all ages, and used more often in non-Hispanic, English-speaking, insured children than other groups (CDC, 2003, Visser et al., 2007). Cognizant of these difficulties, Drappatz and colleagues in 2006 reviewed 845 published articles on ADHD and growth, and concluded that usable data for a meta-analysis review could be extracted from only 22 (2.6%) of published studies (Drappatz et al., 2006).

The pathophysiology of ADHD itself is poorly understood (Spencer et al.,1998). It is likely therefore that the ADHD population in many studies is actually fairly heterogenous, further adding to the problems of patient selection for population studies and the selection of appropriate controls.

Because of the difficulties sited above, the actual long-term growth effects, if any, of ADHD or its treatment have remained unclear and controversial ( Spencer et al.,1998, Pliszka, 1998, Poulton, 2005, Negrao & Viljoen, 2011). Recently, however, long-term, case control data have begun to emerge on final adult heights in ADHD. Biederman and colleagues studied the effect of ADHD and stimulant treatment using longitudinal, case-control studies of male and female children with ADHD (n = 137) compared to control children without ADHD (n = 124) followed for 10–11 years into adulthood. Compared to the control children, there were no differences in any growth outcomes in the ADHD group, regardless of stimulant treatment or the lack thereof (Biederman et al., 2010). A summary of growth studies in ADHD children and adolescents, some with no medical treatment and some treated with psychostimulant medications, is presented in table 2.

| | Reference | Patient Population | Study Details | Result |
|---|---|---|---|---|
| **ADHD, No Stimulant Treatment** | | | | |
| | Lisska & Rivkees, 2003 | ADHD vs. age- and gender-matched, non ADHD siblings, | Height, weight, and BMI prior to ADHD stimulant treatment | No significant differences in height SDS between matched ADHD and non-ADHD siblings |
| | Ptacek et al., 2009a | ADHD males (n=46, ave. age 11.03 years) vs. control, non-ADHD | Anthropometric measurements | Decreased percent body fat and height in ADHD vs. non-ADHD |
| | Biederman et al., 2010 | ADHD males and females (n=137) vs. control, non-ADHD (n=124) | Case-controlled study, longitudinal 10-11 year follow-up | No differences in height trajectories or in any growth parameters at follow-up |
| | Ptacek et al., 2009b | ADHD males (treated and non-treated, n=104, age 4-16 years) vs. control, non-ADHD | Anthropometric measurements | Decreased percent body fat, abdominal circumference, height, head circumference, ADHD vs. non-ADHD |
| **ADHD with Stimulant Treatment** | | | | |
| | Spencer et al.,1998 | ADHD males (89% treated at some time and 11% non-treated, n=124) vs. control, non-ADHD (n=109) | Anthropometric measurements | Mean height deficit of 2.1 cm in early adolescent ADHD males vs. controls. Mean height SDS 0.21 (ADHD) vs. 0.47 (controls), $p = 0.03$ |
| | Lahat et al., 2000 | Methylphenidate treated ADHD vs. control, non ADHD, 1-2 years of treatment | Dual photon absorptiometry, serum alkaline phosphatase, urinary deoxypyridinoline | No differences in bone mineral density turnover in treated ADHD vs. controls |

| | | | | |
|---|---|---|---|---|
| | Lisska & Rivkees, 2003 | Methylphenidate treated ADHD vs. age- and gender-matched, non ADHD siblings, 2-3 years treatment duration | Height, weight and BMI; pre-treatment and up to 3 years during treatment | Decreased height velocity during treatment; overall loss of 0.5 SDS height (boys) and 0.6 SDS height (girls) |
| | Drappatz et al., 2006 | Variable age ranges | Meta-analysis review of 22 studies | 1 - 2 cm / year height deficit per year of ADHD treatment |
| | Poulton, 2005 | Variable age ranges | Medline search and review of 29 studies | One cm / year height deficit, during 1-3 three years of ADHD treatment |
| | Frindik et al., 2009 | Prepubertal, male and female, treated ADHD vs. non - ADHD | Retrospective review of GH registry, pre-GH treatment data | Decreased BMI in treated ADHD; No difference in mean height SDS or growth rates prior to GH treatment |
| | Biederman et al., 2010 | Males and females; psychostimulant treatment ADHD vs. control, non-ADHD | Longitudinal, case-controlled study, linear growth, 10-11 year follow-up | No differences in growth in ADHD with psychostimulant treatment vs. controls |
| | Faraone et al., 2010 | LDX treated ADHD children (n=281), ages 6 to 13 years | Longitudinal, height, weight, BMI, up to 15 months treatment | Decreased gains in height, weight, BMI vs. expected from CDC standards |
| | Rose et al., 2011 | Treated ADHD (n=1055) vs. non - ADHD (n= 6319) | Review of GH registry, pre-GH treatment data | No difference in mean height SDS treated ADHD vs. non-ADHD prior to GH treatment |

Abbreviations: ADHD, attention deficit hyperactivity disorder; n = number, BMI, body mass index; GH, growth hormone; SDS, standard deviation score, LDX = .lisdexamfetamine dimesylate, CDC = Centers for Disease Control.

Table 2. Growth in Children with ADHD

# 4. Growth hormone treatment of ADHD associated short stature

As the preceding section discusses, the majority of children with ADHD have, at worst, mild, perhaps stimulant therapy-related, growth delays, ultimately achieving normal heights and weights. However, a subgroup of ADHD patients seems to exist that does have clinically significant and concerning growth delay. Spencer and colleagues found that 10% of ADHD boys in their review of 124 patients had height SD scores of more than 2 SD's below the mean. Using the same height SD cutoff criteria, only 1% of their control non-ADHD population was this short (Spencer et al.,1998).

The characteristics and natural history of this more growth delayed ADHD subgroup are not well defined. Their documented existence, however, supports our clinical impressions that at least some ADHD-treated children have growth delays that are sufficient to warrant investigation by a pediatric endocrinologist. In point of fact, our pediatric endocrine practice has seen a steadily increasing number of ADHD-treated children referred for evaluation of delayed growth and possible growth hormone therapy. Anecdotally, such children, if treated with recombinant human growth hormone (GH), did not seem to respond as well to GH therapy as children receiving GH therapy alone, suggesting that ADHD treatment might be a relative contraindication to GH therapy.

We undertook a retrospective analysis of children enrolled in a national growth hormone patient registry (Genentech's National Cooperative Growth Study (NCGS)) database (1985-2005) to determine (1) the frequency of ADHD treatment among children who are also receiving GH, (2) if any differences existed in either biochemical testing or anthropometrics between ADHD and non-ADHD, and (3) the first-year growth response in children receiving GH therapy only vs. those receiving GH therapy plus an ADHD medication (Frindik et al., 2009).

## 4.1 Frequency of ADHD medication use in growth hormone treated children

Both the absolute number of children receiving both ADHD treatment and growth hormone and the percentage of such enrolled in the national growth registry increased over a twenty year period. During the first year of the NCGS registry (1985), out of 850 patients receiving GH therapy, only 7 (0.8%) were receiving concurrent ADHD medications. Twenty years later (2005), of 12,113 enrolled patients on GH therapy, 752 (5.8%) were also receiving ADHD treatment. Reasons for this increase were felt to include: 1) an increased incidence of ADHD and/or an increased use of ADHD medication within the general pediatric population (CDC 2003, Froehlich et al., 2007) that could then be reflected in the NCGS database and 2) a referral bias in growth registries (Finkelstein et al., 1998, Kemp, 2006) that possibly leading to some overlap between the GH-treated and the ADHD-treated populations (Visser et al., 2007).

## 4.2 Endocrine function and growth in ADHD children prior to GH treatment

To examine differences in hypothalamic - pituitary - growth hormone axis activity in non-ADHD as compared to ADHD children, children were divided into two groups on the basis of their response to GH stimulation testing. Children were considered to have idiopathic GH deficiency (IGHD) if their maximum stimulated growth hormone (MSGH) response to provocative stimuli was less than 10 ng/ml, or idiopathic short stature (ISS) if the MSGH response was equal to or greater than 10 ng/ml and no other syndrome or diagnosis given. Once divided into these two groups (IGHD and ISS), we found no distinguishing characteristics in hypothalamic - pituitary - growth hormone axis activity: mean MSGH (ng/ml) responses were similar in both the ADHD-treated and non-ADHD-treated groups.

### 4.3 Anthropometrics

The two groups were also similar as regards pre-treatment anthropometric measurements, although the ADHD-treated populations, both IGHD and ISS, were thinner (mean -0.4 BMI standard deviation scores (SDS) for IGHD and -0.4 BMI SDS for ISS) than the non–ADHD-treated, presumably an effect of the ADHD medication on appetite and weight. This difference in BMI was not reflected in height, as there was no statistically significant difference between mean baseline height SDS with and without ADHD treatment. Mean baseline height SDS with ADHD treatment was -2.6 and without ADHD treatment -2.7.

### 4.4 Response to GH treatment

When adjusted for age, sex, and enrollment body mass index, the difference in first-year GH-treatment response for children with IGHD was similar regardless of ADHD therapy. Mean first year growth rate for ADHD-treated IGHD was 8.5±2.0 cm/yr vs 9.4±2.6 cm/yr for non-ADHD-treated IGHD, a slightly less, but clinically insignificant difference. Similarly, first-year growth response was clinically indistinguishable between the ISS groups: 8.1±1.9 cm/yr (ADHD treatment) vs 8.6±2.1 cm/yr (ISS without ADHD treatment).

#### 4.4.1 Other GH registry based studies of GH treatment in ADHD

Rao and colleagues reviewed the impact of ADHD treatment on the response to GH therapy over a longer treatment time, comparing IGHD and ISS children receiving GH plus either methylphenidate and pemoline to the response to GH alone over a mean treatment time of 2.7 to 3.0 years. Enrollment, pre-GH treatment height SD scores were similar in the ADHD treated groups and non-ADHD groups. Treatment with methylphenidate and pemoline had a minor effect on the change in height SDS with GH treatment in the IGHD group. However, not only was this negative effect minor in the first place, but the degree of differential response between the ADHD-IGHD-GH group and the IGHD-GH only group decreased the longer GH treatment was used. No differences were seen in the response to GH therapy in the two ISS groups. Enrollment heights and responses to GH therapy are presented in Table 3.

Another GH registry review of ADHD treated children followed for 3 years of GH therapy also demonstrated a similar response to GH therapy regardless of the status of ADHD treatment.

Children naive to GH therapy were divided into two groups: those receiving concomitant ADHD medication (ADHDm) and those receiving GH only. Diagnoses of patients receiving GH only included GH deficiency, multiple pituitary hormone deficiency, Turner syndrome, Noonan syndrome, small for gestational age, and idiopathic or other short stature. Mean height standard deviation scores (SDS) in the two groups were similar at

| Reference | Patient Population | Treatment Status | Study Parameters | Published Results mean ± SD |
|---|---|---|---|---|
| Rao et al., 1998 | IGHD with treated ADHD vs. non-ADHD IGHD | methylphenidate or pemoline in ADHD | Pre-GH treatment height SDS | - 2.8 ± 0.7 (ADHD) vs. - 3.0 ± 0.9 (non-ADHD) |
| | n=184 (ADHD); n=2313 (IGHD) | methylphenidate or pemoline in ADHD plus GH | Change in height SDS after GH treatment | 1.2 ± 0.8 (ADHD) vs. 1.3 ± 0.9 (non-ADHD) |
| | ISS with treated ADHD vs. non-ADHD ISS | methylphenidate or pemoline in ADHD | Pre-GH treatment height SDS | - 2.8 ± 0.7 (ADHD) vs. - 2.9 ± 0.7 (non-ADHD) |
| | n=117 (ADHD); n=1283 (ISS) | methylphenidate or pemoline in ADHD plus GH | Change in height SDS after GH treatment | 1.0 ± 0.7 (ADHD) vs. 1.1 ± 0.7 (non-ADHD) |
| Frindik et al., 2009 | IGHD with treated ADHD vs. non-ADHD IGHD | stimulant treatment in ADHD | Pre-GH growth rate cm/yr | 3.8 ± 1.8 (ADHD) vs. 4.4 ± 2.3 (non-ADHD) |
| | n= 263 (ADHD); n=6223 (IGHD) | stimulant treatment in ADHD plus GH | Growth Rate cm/yr after one year of GH | 8.5 ± 2.0 (ADHD) vs. 9.4 ± 2.6 (non-ADHD) |
| | ISS with treated ADHD vs. non-ADHD ISS | stimulant treatment in ADHD | Pre-GH growth rate cm/yr | 4.1 ± 1.9 (ADHD) vs. 4.4 ± 2.0 (non-ADHD) |
| | n= 121 (ADHD); n=2695 (ISS) | stimulant treatment in ADHD plus GH | Growth Rate cm/yr after one year of GH | 8.1 ± 1.9 (ADHD) vs. 8.6 ± 2.1 (non-ADHD) |
| Rose et al., 2011 | Patients receiving up to 3 years of GH treatment; n = 314 (ADHD); n = 1583 (GH only) | stimulant treatment in ADHD plus GH vs. non-ADHD with GH only | Change in height SDS after 3 years GH | 1.14 ± 0.60 (ADHD + GH) vs. 1.26 ± 0.79 (non-ADHD) |

Abbreviations: ADHD = attention deficit hyperactivity disorder; IGHD = idiopathic growth hormone deficiency; ISS = idiopathic short stature; n = number; SD = Standard Deviation; SDS = Standard Deviation Score.

Table 3. Effects of Growth Hormone (GH) on Growth in ADHD from Retrospective Reviews of GH Registries

start of therapy: -2.3 ADHDm group vs. -2.2 GH only group. Mean BMI was also comparable between the groups both at enrollment and during treatment. The ADHDm had lower mean change (improvement) in height SDS at 4 months, 1 year, 2 years and 3 years of GH therapy compared to the GH-only group, but the differences were minor and felt to be of no clinical significance. After three years of GH treatment, the change in mean height SDS of the ADHDm group was 1.14 compared to 1.26 of the GH-only group. Furthermore, 83% of patients on GH therapy plus concomitant ADHD medical treatment achieved a height in the normal range (greater than - 2 SDS), a response similar to the non-ADHD, GH-only group in whom 85% of children achieved a height greater than - 2 SDS (Rose et al., 2011).

See Table 3 for a summation of pretreatment data (either enrollment heights or growth velocity, when available) and the effects of GH therapy (either change in height or growth rate) in these three reviews.

## 5. Conclusion

The retrospective nature and referral bias of growth registries (Blethen et al., 1996) prevent extrapolation of these results to the general ADHD population. However, matched, ADHD-treated and non-treated populations are similar enough as regards pre-treatment anthropometric measurements, growth hormone stimulation results, and overall response to GH therapy to support the idea that there is no inherent hypothalamic - pituitary - growth hormone dysfunction in ADHD that would interfere significantly with GH therapy.

Finally, ADHD medical therapy should not exclude GH treatment if a child is otherwise an appropriate candidate for GH therapy. Such children should be expected to respond to GH regardless of concurrent ADHD treatment, many with first-year growth rates very similar to matched, non-ADHD children (Frindik et al., 2009). While it is true that some children with concurrent ADHD treatment may have a slightly decreased response during GH therapy, the effect is minor and of little clinical significance (Rao et al., 1998), and the overwhelming majority ultimately achieve heights in the normal range with GH therapy (Rose et al., 2011).

## 6. References

Anu-Katriina P, Kajantie E, Alexander J, Pyhälä R, Lahti J, Heinonen K, Eriksson JG, Strandberg TE, Räikkönen K. Symptoms of attention deficit hyperactivity disorder in children are associated with cortisol responses to psychosocial stress but not with daily cortisol levels. J Psychiatr Res. 2011 Jul 27. PMID: 21802096

Barzman DH, Patel A, Sonnier L, Strawn JR. Neuroendocrine aspects of pediatric aggression: Can hormone measures be clinically useful? Neuropsychiatr Disease Treatment. 2010 Oct 11;6:691-7. PMID: 21127686

Biederman J, Spencer TJ, Monuteaux MC, Faraone SV. A naturalistic 10-year prospective study of height and weight in children with attention-deficit hyperactivity disorder grown up: sex and treatment effects. Journal Pediatr. 2010 Oct;157(4):635-40, 640.e1. PMID:20605163

Blethen SL, Allen DB, Graves D, August G, Moshang T, Rosenfeld R. Safety of recombinant deoxyribonucleic acid-derived growth hormone: the National Cooperative Growth Study experience. Journal Clin Endocrinol Metab 1996;81:1704–1710.

Centers for Disease Control and Prevention (CDC). Mental health in the United States. Prevalence of diagnosis and medication treatment for attention-deficit/hyperactivity disorder--United States, 2003. MMWR Morbidity Mortality Weekly Report 2005;54:842–847.

Drappatz J, Khwaja O, Neovius M, Sarco D. Growth in children with ADHD treated with stimulant medications: A meta-analysis. Presented at: Pediatric Academic Societies' 2006 Annual Meeting; April 29-May 2, 2006; San Francisco, CA. Abstract 4885.458.

Faraone SV, Biederman J, Morley CP, Spencer TJ. Effect of stimulants on height and weight: a review of the literature. Journal of the American Academy of Childhood and Adolescent Psychiatry. 2008 Sep;47(9):994-1009. PMID:18580502

Faraone SV, Spencer TJ, Kollins SH, Glatt SJ. Effects of lisdexamfetamine dimesylate treatment for ADHD on growth. Journal of the American Academy of Childhood and Adolescent Psychiatry. 2010 Jan;49(1):24-32. PMID:20215923

Finkelstein BS, Silvers JB, Marrero U, Neuhauser D, Cuttler L. Insurance coverage, physician recommendations, and access to emerging treatments: growth hormone therapy for childhood short stature. JAMA 1998;279:663–668.

Freitag CM, Hänig S, Palmason H, Meyer J, Wüst S, Seitz C. Cortisol awakening response in healthy children and children with ADHD: impact of comorbid disorders and psychosocial risk factors. Psychoneuroendocrinology. 2009 Aug;34(7):1019-28. Epub 2009 Mar 10. PMID:19278790

Frindik JP, Fowlkes J, Kemp SF, Thrailkill K, Morales A, Dana K. Stimulant Medication Use and Response to Growth Hormone Therapy: An NCGS Database Analysis. Hormone Research, 72:160–166, 2009.

Froehlich TE, Lanphear BP, Epstein JN, Barbaresi WJ, Katusic SK, Kahn RS. Prevalence, recognition, and treatment of attention-deficit/hyperactivity disorder in a national sample of US children. Arch Pediatr Adolesc Med 2007;161:857–864.

Graham J, Banaschewski T, Buitelaar J, Coghill D, Danckaerts M, Dittmann RW, Döpfner M, Hamilton R, Hollis C, Holtmann M, Hulpke-Wette M, Lecendreux M, Rosenthal E, Rothenberger A, Santosh P, Sergeant J, Simonoff E, Sonuga-Barke E, Wong IC, Zuddas A, Steinhausen HC, Taylor E; European Guidelines Group. European guidelines on managing adverse effects of medication for ADHD. Eur Child Adolesc Psychiatry. 2011 Jan;20(1):17-37. Epub 2010 Nov 3.PMID:21042924

Hastings PD, Fortier I, Utendale WT, Simard LR, Robaey P. Adrenocortical functioning in boys with attention-deficit/hyperactivity disorder: examining subtypes of ADHD and associated comorbid conditions. Journal of Abnormal Child Psychology. 2009 May;37(4):565-78. PMID:19132527

Hatzinger M, Brand S, Perren S, von Wyl A, von Klitzing K, Holsboer-Trachsler E. Hypothalamic - pituitary - adrenocortical (HPA) activity in kindergarten children: importance of gender and associations with behavioral/emotional difficulties. Journal Psychiatr Research. 2007 Nov;41(10):861-70. Epub 2006 Sep 18.PMID:16979188

Jensen JB, Garfinkel BD. Neuroendocrine aspects of attention deficit hyperactivity disorder. Endocrinology and Metabolism Clinics of North America. 1988 Mar;17(1):111-29. Review. PMID: 2897907

Kemp S. Growth hormone deficiency. eMedicine: Pediatrics General Medicine Web site. Arlan L Rosenbloom, editor. Updated April 19, 2006. Available at: http://www.emedicine.com/ped/topic1810.htm.

Lahat E, Weiss M, Ben-Shlomo A, Evans S, Bistritzer T. Bone mineral density and turnover in children with attention-deficit hyperactivity disorder receiving methylphenidate. Journal Child Neurol. 2000. PMID:10921512

Lisska MC, Rivkees SA. Daily Methylphenidate Use Slows the Growth of Children: A Community Based Study. Journal of Pediatric Endocrinology & Metabolism. 2003. 16:711-718.

Ma L, Chen YH, Chen H, Liu YY, Wang YX. The function of hypothalamus - pituitary - adrenal axis in children with ADHD. Brain Research. 2011 Jan 12;1368:159-62. Epub 2010 Nov 12. PMID: 20971091

Negrao BL, Viljoen M. Stimulants and growth in children with attention-deficit / hyperactivity disorder. Medical Hypotheses. 2011 Jul;77(1):21-8. Epub 2011 Mar 27. PMID:21444155

Pliszka SR. The use of psychostimulants in the pediatric patient. Pediatric Clinics of North America. 1998 Oct;45(5):1085-98. Review. PMID: 9884676

Ptacek R, Kuzelova H, Paclt I, Zukov I, Fischer S. Anthropometric changes in non-medicated ADHD boys. Neuro Endocrinology Letters. 2009;30(3):377-81. PMID:19855363

Ptacek R, Kuzelova H, Paclt I, Zukov I, Fischer S. ADHD and growth: anthropometric changes in medicated and non-medicated ADHD boys. Med Sci Monit. 2009 Dec;15(12):CR595-9. PMID:19946228

Rose SR, Reeves G, Gut R, Germak J. Impact of Attention Deficit Hyperactivity Disorder (ADHD) Treatment on Response to Growth Hormone Therapy (GHT). Pediatric Academic Societies (PAS) and Asian Society for Pediatric Research 2011 Annual Meeting: Abstract 1402.24. Presented April 30, 2011.

Spencer T, Biederman J, Wilens T. Growth deficits in children with attention deficit hyperactivity disorder. Pediatrics. 1998 Aug;102(2 Pt 3):501-6. Review. PMID: 9685453

Tobaiqy M, Stewart D, Helms PJ, Williams J, Crum J, Steer C, McLay J. Parental reporting of adverse drug reactions associated with attention-deficit hyperactivity disorder (ADHD) medications in children attending specialist paediatric clinics in the UK. Drug Safety. 2011 Mar 1;34(3):211-9. PMID:21332245

Poulton A. Growth on stimulant medication; clarifying the confusion: a review. Archives Disease Children. 2005 Aug;90(8):801-6. Review. PMID: 16040876

Stadler C, Kroeger A, Weyers P, Grasmann D, Horschinek M, Freitag C, Clement HW. Cortisol reactivity in boys with attention-deficit/hyperactivity disorder and disruptive behavior problems: the impact of callous unemotional traits. Psychiatry Research. 2011 May 15;187(1-2):204-9. Epub 2010 Sep 1. PMID: 20813414

Visser SN, Lesesne CA, Perou R. National estimates and factors associated with medication treatment for childhood attention-deficit/hyperactivity disorder. Pediatrics 2007;119 Suppl 1:S99–106.

# 4

# Attention Deficit Hyperactivity Disorder: Birth Season and Epidemiology

Cristina Morales, Amalia Gordóvil, Jesús Gómez, Teresita Villaseñor, Maribel Peró and Joan Guàrdia

*[1]Centro Universitario de Ciencias de la Salud, Departamento de Neurociencias Universidad de Guadalajara,*
*[2]Departamento de Metodología de las Ciencias del Comportamiento, Facultad de Psicología, Universidad de Barcelona and Institut de Cervell, Cognició i Conducta (IR3C),*
*[1]Mexico*
*[2]Spain*

## 1. Introduction

Attention Deficit Hyperactivity Disorder is a neurodevelopmental disorder that is most frequently diagnosed in the pediatric population. This disorder consists of short-, medium-, and long-term interruptions in development that affect performance and daily activities (Amador et al., 2002). Weiss and Trokenberg (1993) argue that adolescents with Attention Deficit Hyperactivity Disorder show poor scholastic performance, low levels of self-esteem, higher levels of alcohol and drug use, and more difficulties with social functioning when compared to children not diagnosed with Attention Deficit Hyperactivity Disorder.

Given the evidence of difficulties with response inhibition, recent studies have highlighted problems with working memory, attentive faculties (e.g., sustained attention), and response inhibition when faced with interferences (Brito et al., 1999; Déry et al., 1999; Drechsler et al., 2005; Goldberg et al., 2005; Rodríguez et al., 2009; Seidman et al., 2006; Bedard et al., 2007). The fact that these problems appear in childhood, which is a key period for establishing the foundation for future professional exploration, interests, values, attitudes, and vocational abilities (Araújo & Taveira, 2009), conveys the gravity of the situation for a child with Attention Deficit Hyperactivity Disorder.

According to the Secretary of Health in Mexico (2000) and the American Psychiatric Association (2000), the prevalence of Attention Deficit Hyperactivity Disorder is 3-7% in children between 6 and 12 years old. Recent data (World Health Organization, 2005) suggest that the prevalence of Attention Deficit Hyperactivity Disorder in the Mexican child population is 4%, with a girl:boy ratio of 1:5.

A numerous studies have examined risk factors that predispose a child to this disorder. A number of interesting studies have been conducted investigating genetic etiology of Attention Deficit Hyperactivity Disorder (Faraone et al., 1994, 1995; Hudziak et al., 2005; Wallis et al., 2008) that have allowed for the identification of candidate genes associated with the disorder, such as the repetition of allele 7 on the receptor gene for dopamine (Faraone et al., 2001).

Studies on environmental etiology have examined the following risk factors for Attention Deficit Hyperactivity Disorder: childhood exposure to high levels of lead; cranioencephalic trauma affecting the prefrontal cortex; premature birth; low birth weight; maternal consumption of alcohol or tobacco during pregnancy (Anderson & Doyle, 2004; Taylor et al., 2004); young maternal age (Smidt & Osterlam, 2007); situations of psychosocial adversity; and complications during pregnancy (Biederman et al., 1995).

Some studies (Atladóttir et al., 2007; Brookes et al., 2008; Liederman & Flannery, 1994; Mick et al., 1996; Seeger et al., 2004) conducted with pediatric patients suggest that the season of birth may contribute to the later development of Attention Deficit Hyperactivity Disorder.

From a biological perspective, the seasonal hypothesis argues that the season of birth is representative of risk factors, such as viral infections (Mick et al., 1996). For this reason, season of birth is identified as an operational variable in studies analyzing the seasonal effect on the early development of psychopathy. In spite of evidence supporting this phenomenon in schizophrenia (Bradbury & Millar, 1985; D'Amato et al., 1996; Faustman et al., 1992; Franzek & Beckmann, 1992; Tochigi et al., 2004), cerebral tumors (Brenner et al., 2004), and autism (Stevens et al., 2000), the findings regarding Attention Deficit Hyperactivity Disorder remain uncertain.

Atladóttir et al. (2007) examined seasonal variation in the birth of children with autism spectrum disorder, Tourette syndrome, obsessive-compulsive disorder, and Attention Deficit Hyperactivity Disorder. They did not find substantial variations in birth within the different groups. However, they reported evidence of a relationship between season of birth and Attention Deficit Hyperactivity Disorder (i.e., the highest number of births being in autumn and the lowest in spring). Liederman and Flannery (1994) investigated the relationship between neurodevelopmental disorders and season of birth, concluding that being born in spring and summer increased the risk of developing these disorders, including Attention Deficit Hyperactivity Disorder. Following this line of research, Schneider and Eisenberg (2006) also pointed to summer as the season of birth associated with high rates of Attention Deficit Hyperactivity Disorder.

Seeger et al. (2004) suggest that children with a copy of allele DRD4 7 who are born in the spring and summer have a greater risk of developing Attention Deficit Hyperactivity Disorder with conduct disorder. Studies focusing on pediatric patients with Attention Deficit Hyperactivity Disorder as the only diagnosis show a similar relationship between the repeated allele 7 of the DRD4 gene and Attention Deficit Hyperactivity Disorder without the association or interrelation with season of birth.

In a landmark study, Mick et al. (1996) did not find significant differences between the season of birth patterns of patients with Attention Deficit Hyperactivity Disorder and control subjects. Although, they suggested that there was a season of birth pattern with regard to subtypes, in that September births were highly related to diagnoses of Attention Deficit Hyperactivity Disorder with learning disabilities, Attention Deficit Hyperactivity Disorder with psychiatric comorbidity, and Attention Deficit Hyperactivity Disorder with family history of the disorder.

The conclusions regarding season of birth and Attention Deficit Hyperactivity Disorder are divergent, and none of the published studies limited their research to the Latin-American population.

The objective of the present study is to determine if the season of birth effect, which is based on the hypothesis of seasonality that is observed in some psychopathologies, is applicable to children with Attention Deficit Hyperactivity Disorder in the Mexican population. This study will investigate whether the season of birth implies a risk for later development of child Attention Deficit Hyperactivity Disorder. If it does, observing this relationship between season of birth and Attention Deficit Hyperactivity Disorder would allow for the prediction of a determined subtype of the disorder. Following research on schizophrenia and autism, this study will attempt to confirm the relationship between season of birth and child Attention Deficit Hyperactivity Disorder.

The analyses that will be conducted incorporate the variables of age and gender as possible modifiers of this effect. The inclusion of both variables will allow the presentation of epidemiological data regarding child Attention Deficit Hyperactivity Disorder in the Mexican clinical population.

## 2. Method

### 2.1 Subjects

The criteria for inclusion in this sample were the following: a) the age of the patient was between 6 to 12 years, b) the complete information on the patient's birth dates and ages was available, c) the patient's birth place was within the state of Jalisco (Mexico), d) the patient's clinical history was available in the healthcare center, and e) the patient had a suspected diagnosis of Attention Deficit Hyperactivity Disorder.

This study recruited patients being seen during the evening shift in the Neuroscience unit of the Neuroscience Department at the University Center for Health Sciences at the University of Guadalajara (Jalisco, Mexico). The patients were seeking care for behavioral problems or poor scholastic performance, as identified by their parents and/or teachers.

A total of 286 patients between the ages of 6 and 12 years were evaluated; the average age was 8.34 years (SD=.106). Boys accounted for 78.00% (n=223) of the sample, and girls accounted for 22.00% (n=63) of the sample.

Regarding to the distribution of season of birth in the total sample (N=286), 23.80% of patients born in winter, 24.50% born in summer, 25.20% born in autumn and the remaining 26.60% of patients born in spring (see fig. 1).

The diagnostic process indicated that Attention Deficit Hyperactivity Disorder was present in 86.70% (n=248) of the cases.

Patients diagnosed with Attention Deficit Hyperactivity Disorder showed the following subtype distribution: 8.10% for the Hyperactivity/Impulsivity subtype, 27.10% for the Inattentive subtype and the remaining 64.80% for the Combined subtype.

With this sample size, there was a total precision of .058, with a confidence level of 95%, which is below the assumed maximum indeterminacy ($\pi$=.5).

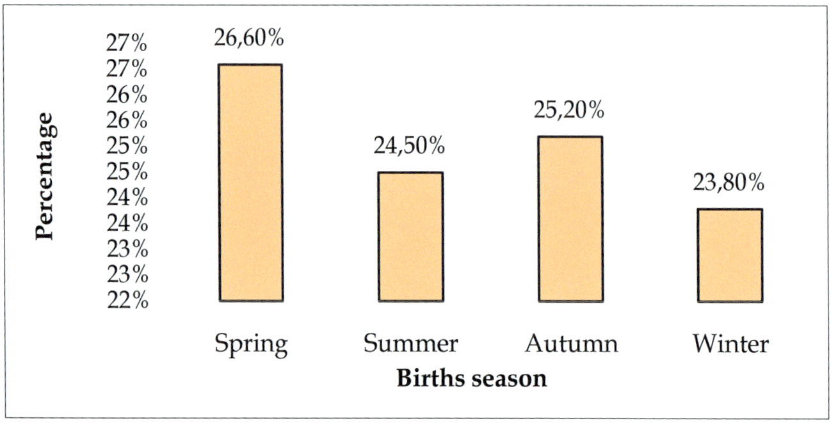


Fig. 1. Season of birth distribution (N=286).

### 2.2 Instruments

Diagnostic determinations were conducted using various sources of information.

First, a retrospective analysis of the patient's clinical history was performed, which included patient data, personal and familial history, the reason for the consultation, and the diagnostic hypothesis.

Complementing this analysis and in compliance with the healthcare centers' protocol, a medical history was taken from the primary caregiver, which included questions regarding perinatal, motor, visual, and auditory development, language and communication, social interaction, independence, and variables related to sleep.

Additionally, a clinical exam of the patient was performed. Finally, the DSM-IV-TR (American Psychiatric Association, 2000) diagnostic criteria for Attention Deficit Hyperactivity Disorder were applied. The information obtained from these sources was used to confirm whether there were positive cases of Attention Deficit Hyperactivity Disorder.

### 2.3 Procedure

Prior to the patient's visit, a review of the corresponding clinical history, which was archived in the healthcare center, was performed. With this information, patient inclusion in the study was determined.

The selected sample underwent a diagnostic process confirming the diagnosis of Attention Deficit Hyperactivity Disorder, which was conducted using medical history interviews with the primary caregiver, as well as through a clinical evaluation of the patient.

Finally, the diagnostic criteria for Attention Deficit Hyperactivity Disorder from the DSM-IV-TR (American Psychiatric Association, 2000) were applied, and patients were classified according to whether they had Attention Deficit Hyperactivity Disorder or not, and Attention Deficit Hyperactivity Disorder subtypes (i.e., hyperactive/impulsive, inattentive, or combined). The interview, clinical evaluation, and application of diagnostic criteria were completed in one session.

A team of two doctors and two neuropsychologists performed all of the evaluations. All of the members of the team were trained in this process, and empirical evidence showed sufficient concordant criteria conditions between evaluators ($\varphi = .86$).

All of the primary caregivers were informed of the confidentiality of the data, and informed consent was granted for participation in the study. The evaluations were performed between January 2001 and February 2006.

### 2.4 Data analysis

We performed a preliminary phase of univariate analysis to detect possible anomalies in the distribution of the variables.

This phase was followed by the description of variables (i.e., season of birth, age, gender, presence of Attention Deficit Hyperactivity Disorder and Attention Deficit Hyperactivity Disorder subtypes).

Then, goodness-of-fit tests to analyze the distribution of births in the different seasons of the year and bivariate tests of independence were performed, examining associations of interest that would support the subsequent interpretation of the logistic model.

Binary and multinomial logistic regression models were used. The possible classifications for the diagnosis of Attention Deficit Hyperactivity Disorder (i.e., binary model) and its subtypes (i.e., multinomial model) were studied.

Finally, we decided to focus on the following question: is there a seasonal difference in ADHD? We included gender on the analyses in order to analyze possible differences into the season of born between boys and girls diagnosed with Attention Deficit Hyperactivity Disorder.

The analyses were performed with the program PASW Statistics 17 using the *enter* method. An initial set $\alpha$ value of .05 was adjusted "a posteriori" in accordance with Bonferroni to a value of .03.

## 3. Results

Table 1 presents the observed distributions of gender and the season of birth according to the presence of Attention Deficit Hyperactivity Disorder and the observed distribution by subtypes of the disorder in positive cases.

The goodness-of-fit test, which evaluated the distribution of births in the four seasons of the year for the whole sample, was not significant ($\chi^2$=.490; d.f.=3; p=.921). Significant differences regarding the relationship between the presence of Attention Deficit Hyperactivity Disorder and the season of birth were not observed ($\chi^2$=1.281; d.f.=3; p=.734). Therefore, this first bivariate analysis did not show empirical evidence supporting the hypothesis of seasonality.

| | | Attention Deficit Hyperactivity Disorder diagnosis | |
|---|---|---|---|
| | | Yes | No |
| Gender | Boys (n=223) | 89.69% | 10.31% |
| | Girls (n=63) | 76.19% | 23.81% |
| Season of birth | Spring (n=76) | 85.53% | 14.47% |
| | Summer (n=70) | 90.00% | 10.00% |
| | Autumn (n=72) | 87,50% | 12,50% |
| | Winter (n=68) | 83.82% | 16,18% |
| Attention Deficit Hyperactivity Disorder Subtype (n = 248) | Combined | 64.78% | |
| | Hyperactive-Impulsive | 27.12% | |
| | Inattentive | 8.10% | |

Table 1. Observed distribution of gender and season of birth according to presence or absence of Attention Deficit Hyperactivity Disorder and observed distribution of subtypes of the disorder.

Following Seeger et al. (2004), the season of birth was recoded according to whether the photoperiod of the pregnancy was long (i.e., births in autumn or winter) or short (i.e., spring and summer). The relationship between the photoperiod of the pregnancy and the presence of Attention Deficit Hyperactivity Disorder was not statistically significant ($\chi^2$=.238; d.f.=1; p=.728). Given that the climate of the state of Jalisco is characterized by the presence and absence of rain, seasonality was defined in these terms to analyze its relationship with the presence of Attention Deficit Hyperactivity Disorder and to assess if the results are in accord with the two earlier analyses. The analysis of the relationship between the presence of the disorder and the season dichotomized by the presence or absence of rain proved to be equally insignificant ($\chi^2$=.999; d.f.=1; p=.318).

There was statistical significance in the relationship between gender and Attention Deficit Hyperactivity Disorder ($\chi^2$=7.765; d.f.=1; p=.005; $\varphi$=.459) and between gender and Attention Deficit Hyperactivity Disorder subtypes ($\chi^2$=7.423; d.f.=1; p=.006). The relationship between gender and Attention Deficit Hyperactivity Disorder subtypes was moderate ($\varphi$=.44) with fewer boys in the Hyperactive-Impulsive and Inattentive subtypes. There were also more girls than anticipated classified in these subtypes. The combined subtype was most clearly linked to gender, as it was most often diagnosed in boys.

Given the significance of gender in these early analyses, a stricter examination of the effect of gender was performed using the stratified estimation of the odds ratio.

In this analysis, the Mantel-Haenszel statistic proved significant ($\chi_{M\text{-}H}$ = 6.615; d.f.= 1; p = .010). The results are presented in Table 2. As is shown in this table, there is a significant relationship between gender and Attention Deficit Hyperactivity Disorder, with the likelihood ratio of females to males ranging between 1.319 and 5.598. With regard to the absence of the disorder, there was a considerable advantage in favor of the girls (CI to 95%=1.283-4.153), indicating a greater number of girls in the group without Attention Deficit Hyperactivity Disorder than boys. This suggests that being a female acts as a protective factor against the presence of Attention Deficit Hyperactivity Disorder. In the Attention Deficit Hyperactivity Disorder group, a tendency toward significance was found, with a greater number of boys being diagnosed with Attention Deficit Hyperactivity Disorder than girls.

| Source | Value | CI (95%) |
|---|---|---|
| Odds ratio for gender (girls/boys) | 2.717 | 1.319-5.598 |
| Attention Deficit Hyperactivity Disorder diagnosis=No | 2.308 | 1.283-4.153 |
| Attention Deficit Hyperactivity Disorder diagnosis= Yes | .850 | .735-.982 |

CI: confidence interval at a 95% level.

Table 2. Analysis of the odds ratio stratified by gender and presence of Attention Deficit Hyperactivity Disorder.

In the quantitative variables analysis, a possible effect between age and Attention Deficit Hyperactivity Disorder was studied. The average age of participants diagnosed with Attention Deficit Hyperactivity Disorder ($n_1$= 248) was 8.28 years (SD=1.768), and the average age of patients with a negative diagnosis ($n_2$=38) was 8.67 years (SD=1.979). The difference between the averages was not significant (t=1.549; d.f.=284; p=.122); thus, differences between age and final diagnosis (i.e., the positive or negative presence of Attention Deficit Hyperactivity Disorder) are not evident.

As in the earlier analysis, the possible effect of age of girls and boys with Attention Deficit Hyperactivity Disorder was studied. The average age for boys with a positive diagnosis ($n_1$=200) was 8.27 years (SD=1.767). In the group of girls diagnosed with Attention Deficit Hyperactivity Disorder ($n_2$=48), the average age was 8.31 years (SD=1.788). The difference between the average age of the groups was not significant (t=.149; d.f.=246; p=.881).

Binary and multinomial logistic regression models were analyzed.

The goal of the binary logistic model was to determine if a subset of variables was a good predictor of the presence of child Attention Deficit Hyperactivity Disorder. The variables of season of birth, age, and gender were included in this model. Although earlier bivariate analyses did not indicate the significance of season of birth and age, it was necessary to include these variables in this model to observe the aggregate behavior of the selected variables.

The reference categories established for gender and season of birth were female gender (World Health Organization, 2005) and winter, respectively. This season was identified due to it being the only one for which there was not a reported increase in birthrate according to previous studies examining the relationship between Attention Deficit Hyperactivity Disorder and season of birth (Atladóttir et al., 2007; Liederman & Flannery, 1994; Mick et al., 1996; Seeger et al., 2004).

The values for the binary logistic model are shown in table 3.

Statistical significance was reached for the gender variable ($\beta$=1.035; d.f.= 1; p=.006), suggesting that being male allowed for a classification prediction for the patients in the positive case group.

These results obtained through the use of the logistic regression model confirmed the values found in the odds ratio analysis, in which gender was found to be close to being statistically significant (p=.054).

As was indicated previously, a multinomial logistic regression model was used to analyze whether some of the variables allowed for a correct classification prediction for patients diagnosed with Attention Deficit Hyperactivity Disorder of a determined subtype. This analysis indicated that none of the variables introduced in the model reached statistical significance.

| Source | Coefficient | Standard Error | Wald Test | d.f. | P-value | OR=$e^{\beta}$ | CI (95%) |
|---|---|---|---|---|---|---|---|
| Constant | 2.742 | .993 | 7.624 | 1 | .006 | 15.512 | * |
| Season | | | | | | | |
| Season 1 | -.405 | .504 | .645 | 1 | .422 | .667 | .249-1.791 |
| Season 2 | .015 | .553 | .001 | 1 | .979 | 1.015 | .344-2.998 |
| Season 3 | -.562 | .509 | 1.217 | 1 | .270 | .570 | .210-1.547 |
| Gender | 1.035 | .377 | 7.555 | 1 | .006 | 2.816 | 1.346-5.893 |
| Age | -.158 | .098 | 2.615 | 1 | .106 | .854 | .705-1.034 |

d.f. degrees of freedom, OR: odds ratio and CI: confidence interval at a 95% level.
Season1: Spring-Winter
Season 2: Summer-Winter
Season 3: Autumn-Winter

Table 3. Results of estimates of the coefficients of binary logistic regression for predicting the diagnosis of Attention Deficit Hyperactivity Disorder.

Finally, a chi-square test was performed in order to focus on the following question: is there a seasonal difference in ADHD girls and boys? No statistically differences were found between the season of born of girls and boys with an Attention Deficit Hyperactivity Disorder ($\chi^2$=1.099; d.f.=3; p=.777). Hence, there was not a seasonal difference in Attention Deficit Hyperactivity Disorder.

## 4. Conclusion

This study's objective (i.e., the exploration of the effect of seasonality on child Attention Deficit Hyperactivity Disorder) lead to the use of an analytic model with a reduced number of variables. This provided epidemiological data regarding Attention Deficit Hyperactivity Disorder in Jalisco pediatric patients.

The most relevant result was the greater number of positive cases of child Attention Deficit Hyperactivity Disorder among the clinical patients who met the criteria for inclusion (86.7%).

With regard to gender, an increased presence of Attention Deficit Hyperactivity Disorder in boys was observed, which corresponds to existing theories (American Psychiatric Association, 2000; Froehlich et al., 2008; World Health Organization, 2005). From an epidemiological perspective, being female emerges as a factor that protects against the development of Attention Deficit Hyperactivity Disorder.

The investigation of Attention Deficit Hyperactivity Disorder subtypes indicated a higher number of positive cases in the combined subtype, followed by the hyperactive-impulsive subtype, whereas the inattentive subtype was the least prevalent, which corresponds to findings by Cormier (2008). In accord with results presented by Froehlich et al. (2008), the combined subtype appeared to be clearly linked to gender, in that the number of boys diagnosed with Attention Deficit Hyperactivity Disorder combined subtype was significantly greater than the number of girls diagnosed with this subtype. The two remaining subtypes did not appear to be linked to gender. This finding is consistent with an American Psychiatric Association (2000) report that the inattentive subtype is the least likely to be linked with the gender of a patient compared with the other subtypes.

The absence of a significant relationship between age and diagnosis did not provide insight into the determining variables that can help to predict an Attention Deficit Hyperactivity Disorder diagnosis with greater validity.

Given the controversial over-diagnosis of this disorder, it would be of great clinical value to know whether certain variables provide better diagnostic certainty (for example, to know whether a diagnosis is more trustworthy at a particular age). The existing scientific literature highlights the imprecision of evaluations conducted with preschool children (American Psychiatric Association, 2000; Smidt & Osterlam, 2007), although we cannot verify this because preschoolers were excluded from the present study for that very reason.

Although not statistically significant, the average age for girls with Attention Deficit Hyperactivity Disorder was slightly higher than the average age for boys, which suggests that the negative diagnoses in girls' early years does not exclude Attention Deficit Hyperactivity Disorder development in later years. This finding contributes to the stipulated importance of clinical follow-up studies.

In addition to the epidemiological data provided, it is possible to offer some considerations regarding the principal objective of this study: the exploration of a possible seasonal effect on the later development of child Attention Deficit Hyperactivity Disorder. The results indicated the absence of a seasonal effect for this sample of Jalisco pediatric patients.

Similar negative results were found regarding the four seasons of the year, the duration of the photoperiod of the pregnancy, and the climatic dichotomy in this area (e.g., it is characterized by the presence or absence of rain). In the present study, we could not determine if the effect of season of birth on Attention Deficit Hyperactivity Disorder was not plausible in general or if it was only not plausible in the area that was studied.

This nuance is relevant given that the results could be explained by a particular climatic homogeneity in the manner discussed by D'Amato et al. (1996). That is to say, the seasonal variable categories are not mutually exclusive. With this statement, we are not referring to a methodological artifact derived from categorization but, rather, to a seasonal climate in the studied region that has little variation.

Another possible explanation is based on the possibility that the seasonal effect is modulated by other variables and does not present itself in a direct manner, which is how we approached it. Future research should assess the possibility of including variables, like genetic type. That is to say, it should assess the seasonal effect on subjects who present determined genetic characteristics that have an empirically proven relationship to Attention Deficit Hyperactivity Disorder.

The present study does not provide evidence supporting a seasonal effect on child Attention Deficit Hyperactivity Disorder in the Jalisco population. Although some of the data is convincing, future studies that introduce other mediating variables or that are conducted in other regions will provide more information on this topic.

## 5. Acknowledgment

A special note of thanks is due to the *Department* at the *University Center for Health Sciences* at the *University of Guadalajara* (Jalisco, Mexico). for providing access to the study sample. It must also be said that this research has been made possible by a fellowship of the *Comission for Universities and Research from the Generalitat de Catalunya's Department of Innovation, Universities and Enterprise.*

## 6. References

Amador, J.A., Idiázabal, M.A., Sangorrín, J., Espadaler, J.M. & Forns, M. (2002). Utilidad de las escalas de Conners para discriminar entre sujetos con y sin trastorno por déficit de atención con hiperactividad. *Psicothema,* Vol.14, pp. 350-356.

American Psychiatric Association (2000). *Diagnostic and statistical manual of mental disorders: DSM-IV-TR.* (4th Ed), ISBN 0890420254, Washington, D.C, United States of America

Anderson, P.I. & Doyle, L.W. (2004). Executive functioning in school-aged children who were born very preterm or with extremely low birth weight in the 1990s. *Pediatrics,* Vol.114, No.1, (July 2004), pp. 50-57.

Araújo, A.M., & Taveira, M.C. (2009). Estudio del desarrollo de la orientación vocacional en la infancia desde la perspectiva evolutivo-contextual. *European Journal of Education and Psychology,* Vol.2, No.1, pp. 49-67, ISSN 1888-8992

Atladóttir, H.O., Parner, E.T., Schendel, D., Dalsgaard, S., Thomsen, P.H. & Thorsen, P. (2007). Variation in incidence of neurodevelopmental disorders with season of birth. *Epidemiology,* Vol.18, No.2, (March 2007), pp. 240-245.

Bedard, A.C., Jain, U., Hogg-Johnson, S. & Tannock, R. (2007). Effects of methylphenidate on working memory components: influence of measurement. *Journal of Child Psychology and Psychiatry,* Vol.48, No.9, (September 2007) pp. 872-880.

Biederman, L., Milberger, S., Faraone, S.V., Kiely, K., Guite, J., Mick, E., Ablon, S., Warburton, R. & Reed, E. (1995). Family-environment risk factors for attention-deficit hyperactivity disorder. A test of Rutter's indicators of adversity. *Archives of General Psychiatry,* Vol.52, No.6, (June 1995), pp. 464-470.

Bradbury, T. & Miller, G.A. (1985), Season of birth in schizophrenia: a review of evidence, methodology, and etiology. *Psychological Bulletin,* Vol.98, No.3, (November 1985), pp. 569-594.

Brenner, A.V., Linet, M.S., Shapiro, W.R., Selker, R.G., Fine, H.A., Black, P.M. & Inskip, P.D. (2004).Season of birth and risk of brain tumors in adults. *Neurology.* Vol.63, No.2, (July 2004), pp. 276-281.

Brito, G., Pereira, C. & Santos-Morales, T. (1999). Behavioral and neuropsychological correlates of hyperactivity and inattention in Brazilian school children. *Developmental Medicine & Child Neurology,* Vol.41, No.11, (November, 1999), pp. 732-739.

Brookes, K.J., Neale, B., Xu, X., Thapar, A., Gill, M., Langley, K., Hawi, Z., Mill, J., Taylor, E., Franke, B., Chen, W., Ebstein, R., Buitelaar, J., Banaschewski, T., Sonuga-Barke, E., Eisenberg, J., Manor, I., Miranda, A., Oades, R.D., Roeyers, H., Rothenberger, A., Sergeant, J., Steinhausen, H.C., Faraone, S.V. & Asherson, P. (2008). Differential dopamine receptor D4 allele association with Attention Deficit Hyperactivity Disorder dependent of proband season of birth. *American Journal of Medical Genetics Part B,* Vol.147B, No.1, (January 2008), pp. 94-99.

Cormier, E. (2008). Attention Deficit/Hyperactivity Disorder: a review and update. *Journal of Pediatric Nursing,* Vol.23, No.5, (October 2008), pp. 345-357.

D'Amato, T., Guillaud-Bataille, J.M., Rochet, T., Jay, M., Mercier, C., Terra, J.L. & Daléry, J. (1996). No season-of-birth effect in schizophrenic patients from a tropical island in the Southern Hemisphere. *Psychiatry Research,* Vol.60, (March 1996), pp. 205-210.

Déry, M., Toupin, J., Pauzé, R., Mercier, H. & Fortin, L, (1999). Neuropsychological characteristics of adolescents with conduct disorder: association with Attention-Deficit-Hyperactivity and aggression. *Journal of Abnormal Child Psychology,* Vol.27, No.3, (June 1999), pp. 225-236.

Drechsler, R., Brandeis, D., Foldény, M., Imhof, K. & Steinhausen,H.C. (2005). The course of neuropsychological functions in children with Attention Deficit Hyperactivity Disorder from late childhood to early adolescence. *Journal of Child Psychology and Psychiatry,* Vol.46, No.8, (August 2005), pp. 824-836.

Faraone, S.V., Biederman, J., Chen, W.J., Milberger, S., Warburton, R.M. & Tsuang, M.T. (1995). Genetic heterogeneity in attention deficit hyperactivity disorder: gender, psychiatric comorbidity and maternal Attention Deficit Hyperactivity Disorder. *Journal of Abnormal Psychology,* Vol.104, No.2, (May 1995), pp. 334-345.

Faraone, S., Biederman, J. & Milberger, S. (1994). An exploratory study of Attention Deficit Hyperactivity Disorder among second-degree relatives of Attention Deficit Hyperactivity Disorder children. *Biological Psychiatry,* Vol.35, No.6, (March 1994), pp. 398-402.

Faraone, S.V., Doyle, A.E., Mick, E. & Biederman, J. (2001). Meta-Analysis of the association between the 7-repeat allele of the dopamine D4 receptor gene and Attention Deficit Hyperactivity Disorder. *The American journal of psychiatry,* Vol.158, No.7, (July 2001), pp. 1052-1057.

Faustman, W.O., Bono, M.A., Moses, J.A. & Csernansky, J.G. (1992), Season of birth and neuropsychological impairment in schizophrenia. *The Journal of Nervous and Mental Disease,* Vol.180, No.10, (October 1992), pp. 644-648.

Franzek, E. & Beckmann, H. (1992). Season-of-birth effect reveals the existence of etiologically different groups of schizophrenia. *Biological Psychiatry,* Vol.32, No.4, (August 1992), pp. 375-378.

Froehlich, T.E., Lanphear, B.P., Epstein, J.N., Barbaresi, W.J., Katusik, S.K. & Kahn, R.S. (2008). Prevalence, recognition, and treatment of Attention-Deficit/Hyperactivity Disorder in a national sample of US children. *Archives of Pediatrics & Adolescent Medicine,* Vol.161, No.9, (September 2007), pp. 857-864.

Goldberg, M.C., Mostofsky, S.H., Cutting, L.E., Mahone, E.M., Astor, B.C., Denkla, M.B. & Landa, R.J. (2005). Subtle executive impairment in children with Autism and children with Attention Deficit Hyperactivity Disorder. *Journal of Autism and Developmental Disorders,* Vol.35, (June 2005), pp. 279-293.

Hudziak, J.J., Derks, E.M., Althoff, R.R., Rettew, D.C. & Boomsma, D.I. (2005). The genetic and environmental contributions to Attention Deficit Hyperactivity Disorder as measured by the Conners' Rating Scales-Revised. *The American Journal of Psychiatry,* Vol.162, (September 2005), pp. 1614-1620.

Liederman, J. & Flannery, K.A. (1994). Fall conception increases the risk of neurodevelopmental disorder in offspring. *Journal of Clinical and Experimental Neuropsychology,* Vol.16, No.5, (October 1994), pp. 754-768.

Mick, E., Biederman, J. & Faraone, S.V. (1996). Is season of birth a risk factor for attention-deficit hyperactivity disorder? *Journal of the American Academy of Child and Adolescent Psychiatry,* Vol.35, No.11, (November 1996), pp. 1470-1476.

Rodríguez, C., Álvarez, D., González-Castro, P., García, J.N., Álvarez, L., Núñez, J.C., González, J.A. & Bernardo, A. (2009). TDAH y dificultades de aprendizaje en escritura: comorbilidad en base a la atención y memoria operativa. *European Journal of Education and Psychology,* Vol.2, No.3, pp. 181-198 ISSN 1888-8992

Schneider, H. & Eisenberg, D. (2006). Who Receives a Diagnosis of Attention-Deficit/Hyperactivity Disorder in the United States Elementary School Population? *Pediatrics,* Vol.117, No.4, (April 2006), pp. 601-609.

Secretary of Health in México. (2000). Available from www.salud.gob.mx/ssa_app/noticias/datos/2000-12-02_356.html

Seeger, G., Schloss, P., Schmidt, M.H., Rüter-Jungfleisch, A. & Henn, F.A. (2004). Gene-environment interaction in hyperkinetic conduct disorder (HD + CD) as indicated by season of birth variations in dopamine receptor (DRD4) gene polymorphism. *Neuroscience Letters,* Vol.366, No.3, (August 2004), , pp. 282-286.

Seidaman, L.J., Biederman, J., Valera, E.M., Monteaux, M.C., Doyle, A.E. & Faraone, S.V. (2006). Neuropsychological functioning in girls with Attention-Deficit/Hyperactivity Disorder with and without learning disabilities. *Neuropsychology,* Vol.20, No.2, (March 2006), pp. 166-177.

Smidt, D.P. & Osterlam, J. (2007). How common are symptoms of Attention Deficit Hyperactivity Disorder in typically developing preschoolers? A study on prevalence rates and prenatal/demographic risk factors. *Cortex,* Vol.43, No.6, (August 2007), pp. 710-717.

Stevens, M.C., Fein, D. & Waterhouse, L.H. (2000). Season of births effects in autism. *Journal of Clinical and Experimental Neuropsychology,* Vol.22, No.3, (June 2000), pp. 399-407.

Taylor, E., Dopfner, M., Asherson, P., Banaschewski, T., Buitelaar, J., Coghill, D., Danckaerts, M., Rothenberger, A., Sonuga-Barke, E., Steinhausen, H.C. & Zuddas, A. (2004). European clinical guidelines for hyperkinetic disorder-firs upgrade. *European Child & Adolescent Psychiatry,* Vol.13, pp. 7-30.

Tochigi, M., Okazaki, Y., Kato, N. & Sasaki, T. (2004). What causes seasonality of birth in schizofrenia? *Neuroscience Research,* Vol.48, No.1, (January 2004), pp. 1-11.

Wallis, D., Russell, H.F. & Muenke, M. (2008). Review: Genetics of Attention Deficit/Hyperactivity Disorder. *Journal of Pediatric Psychology,* Vol.33, No.10, (December 2008), pp. 1085-1099.

Weiss, G. & Trokenberg, L. (1993). *Hyperactive children grown up: Attention Deficit Hyperactivity Disorder in children, adolescents and adults* (2nd Ed.), Guilford Press, ISBN 0898625963. New York: United states of America.

World Health Organization. (2005). Statistical Database of Health. Available from www.oms.org

5

# Hypothyroxinemia in Pregnancy is Related with Attention Deficit Hyperactivity Disorder

Miriam Muñoz Lopez
*Instituto Clínico de Medicina Materno Fetal (Clinic-Maternitat)*
*Universidad de Barcelona,*
*Spain*

## 1. Introduction

As previously published material has demonstrated, there is a prevalence of 70% of ADHD in children with thyroid hormone syndrome (GRTH)[1], a disorder caused by a mutation in the gene of the thyroid receptor B and characterized by a lowered response to the action of thyroid hormones at pituitary and peripheral tissue levels. Vermiglio[2] have conducted a study that relates the maternal hypothyroxinemia in different trimesters of pregnancy for the development of the child, followed by up to 10 years age. The most important finding, and totally unexpected, was that 70% of the offspring of mothers, with mild iododeficiency had ADHD, while it is not diagnosed in offspring of mothers in a control area without iodo insufficiency . The similarity of the results of these studies would appear to point out that there could be a potential relationship between the neurophysiologic disorder, the GRTH syndrome and the impaired thyroid hormone action. This could be either because of low levels of maternal thyroxine which are due to an insufficient previous iodine intake before and during pregnancy or a defect in the thyroid hormone receptor.

When iodine deficiency is present during the earlier stages of pregnancy and early brain development, leads to neurological cretinism. Throughout gestation, the maternal thyroxine (T4) is transferred to the fetus through the placenta and has a neuroprotective role. The free T4 in the fetal fluids increases in parallel to the maternal T4, and therefore a normal maternal thyroxinemia is of utmost importance for the protection of the fetal brain.

In this chapter we discuss the thyroid hormones and their influence during fetal brain development, by studying the changes covered by these hormones during fetal development and the influence of maternal thyroid function for proper fetal brain development

Thyroid hormones (TH) are secreted mostly as thyroxine (T4) from the thyroid gland and transformed in diverse tissues to the transcriptionally active form 3,5,3′ -triiodothyronine (T3) to the deiodinases type 1 and 2 (D1,D2) (Yen,2001). T3 plays a key role during the central nervous' system (CNS) development. During development, T3 is needed for the normal expression of genes which are critically involved in many processes, especially in neuronal migration and differentiation. Deficiency during the fetal and early postnatal period leads to striking abnormalities in dendritic an axonal growth, synaptognenesis, neuronal migration, myelination and also neuronal cell death (Chan and Kilby, 2000). T4 treatment immediately after birth is enough to prevent most of the brain damage induced by neonatal hypothyroidism. Such treatment, however, cannot fully rescue abnormal brain development, induced by hypothyroidism in the uterus (i.e., by maternal iodine deficiency) (Koibuchi and Iwasaki, 2006).

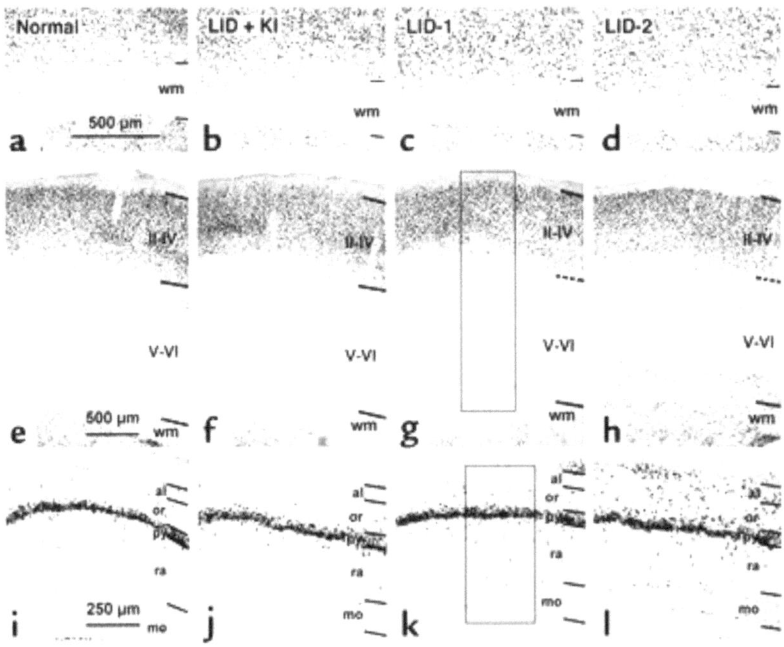


3

Fig. 1. Cortical cell migration in rats induced by a low iodine diet

T3 regulates TRH expression in developing hypothalamic neurons in vitro. The supply, needs a proper concentration of T3 for a normal neuronal development. The data that A. Carreón-Rodriguez et al 2009 presented suggested that T3 plays a key role not only in the HPT axis function but also in its development. Its mechanism is unknown; therefore, an adequate balance of time and concentration of TR isoforms may determine the set point for TRH expression to consequently regulate thyroid hormone levels. An imbalance during the development of the neuroendocrine TR isoforms could result in neurological diseases (ADHD, obesity, anorexia nervosa, subclinical hypothyroidism, etc. (Siesser et al., 2005, 2006; Reinehr et al, 2008)

### 1.1 Pregnancy and thyroid function

Hormonal and metabolic changes during pregnancy result in profound alterations of the biochemical parameters of the thyroid function (D Glioner 2001). In regard to this, the main event that can occur during pregnancy is a marked increase in serum thyroxin-binding globulin levels. This change will require an increased hormonal output by the maternal thyroid gland. The metabolic adjustment will be unable to be reached when the functional capacity of the thyroid gland is impaired because of iodine deficiency or (TAI) thyroid autoimmunity

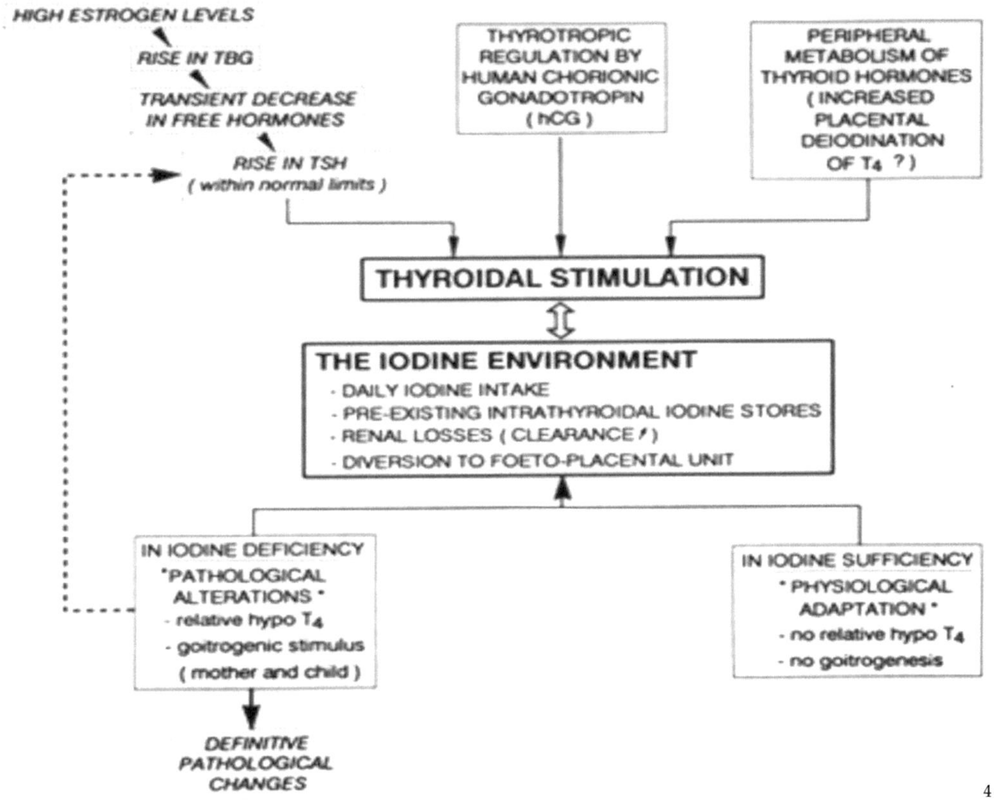


Fig. 2.

### 1.2 Fetal thyroid physiology

Fetal thyroid function starts to occur at the end of the first trimester. Before that, there is evidence that the normal development of the fetal brain is dependent upon maternally derived T4, which is converted intracellularly to T3. This T4 starts to be detected from the 5 to.8 gestation week and by the time pregnancy reaches 11 weeks it is 100 times more concentrated than in the maternal circulation. Maternal hypothyroxinaemia at this stage may have adverse effects on subsequent fetal brain development (J C Girling 2006)

Fetal fT4 and total T4 reach adult levels by 36 weeks' gestation, fetal TSH is greater than adult TSH and fetal T3 remains low. The relatively high levels of T4 allow intracellular conversion to T3 in the fetal brain. So the maternally derived T4 becomes essential to allow normal neurological development. Placental perfusion studies have demonstrated that, in a normal pregnancy very little (0.008%) maternal T4 crosses to the fetal side; inhibition of placental diodination of T4 enhances transfer 2700-fold so that fetal levels reach 30% of maternal concentrations. In a pregnancy complicated by fetal thyroid dysfunction deiodinase III is inhibited, allowing additional transfer of t4 to the fetus reducing fetal peripheral deiodination of t4 and enhancing intracellular activation of T3 in the fetal brain, protecting it from permanent damage.

### 1.3 Iodine deficiency

Iodine deficiency results in cretinism in newborns, goiter in adults and reduced reproductive success in women. More than one billion people worldwide are affected by iodine deficiency and more than 20 million people have adverse neurological sequels that followed fetal iodine deprivation. Neurological endemic cretinism is the leading preventable cause of mental retardation and has a major negative effect on the economy of afflicted areas. In iodinedeficient populations, 2-10% of individuals are affected; in these populations, a mild degree of mental retardation is five times more common than cretinism itself, resulting in a left shift of the intelligence quotient (IQ) distribution curve by 10 points. Cretinism is characterized by deaf mutism, intellectual deficiency, spastic motor disorder and, in some cases, hypothyroidism.

### 1.4 Thyroid autoimmunity

There is increasing evidence that mainly nutritive factors and environmental pollution by metals and chemicals (organ chlorines, pesticides) are the main factors in the present-day spread of this disease. TAI (thyroid autoimmunity) nowadays affects at least the 10% of the population of North America and Europe[5]

It can therefore be assumed that the role of genetic and environmental factors is of utmost importance for the triggering of the disease, with the susceptibility genes contributing by an 80% and environmental factors by a 20% (Tomer Y et al 2009)

The link between nutrigenomics, proteomics and metabolomics is likely to provide a platform for prevention and enhanced treatment of TAI.

It has been reported that certain epigenetic factors are dependent on specific food components such as metals and pollutants increase an individual's susceptibility to type 1 diabetes mellitus or Crohn's disease. These findings unmask a pathogenetic association [6]

It is thus apparent that the environment may play a decisive role in modifying genetic expression via epigenetic mechanism by triggering autoimmunity[7]

### 1.5 Pollutants

Recently, a study in Brazil assessing the prevalence of HT and thyroid antibodies in residents of an area surrounding a petrochemical complex reported a higher incidence of HT (9,3% and anti-thyroid antibodies than in the center area of Sao Paulo)[8]

There is also increasing concern regarding a possible linkage between chronic exposure to both solvents (benzene) and air pollutants (carbon monoxide) with thyroid and respiratory functioning. A recent study that investigated the long-term effects of the exposure to benzene and carbon monoxide in a group of non-smoking petrol station attendants marked increase in T4 and fT4 and decrease in TSH and T3 serum levels[9]

A hypothesis to this was that the organochlorines (OC) body burdens in the young adults were in fact a consequence to the exposure to high levels of OC by their mothers, during the offspring's prenatal and perinatal life[10]

Another study undertaken in East Slovakia showed that fish from waters in the surrounding region consumed by the local population were becoming the prime source of ingested PCBs and pesticides, this leading to a rising incidence of goiter an autoimmunity[11].

What causes this to happen still remains unclear. However, in an experimental study that researched the thyroid function in Sparague-Dawley rats after exposure to Aoclor (PCB s group), distinct histopathological changes such as hyperplasia of epithelia in follicles, reduction of colloid content and lymphocytic infiltration into the perifollicular areas, were observed [12] .In parallel, the decreased FT3 and fT4 and the increased TSH serum concentrations and TPOAB titers indicated that PCBs affect thyroid functioning by inducing autoimmunity.

### 1.6 Smoking

Tobacco smoke contains several potent goitrogens, which interfere with the NIS (cotransportador $NA/I^+$), TPO and dual oxidase (DUOX) activities, rendering smoking a major risk factor for thyroid disease.

Smoking may cause the development of thyroid-associated opthalmopathy in patients with Grave's disease. Smokers receiving radioiodine have the highest incidence of deterioration or recurrence of this disorder[13]

Second-hand smoke (SHS) exposure disrupts thyroid function and induces inflammatory stress by increasing IL-1beta with impairs thyroid hormone synthesis and iodine uptake[14].

It is evident that, as SHS exposure disrupts human biological systems via thyroid impairment, preventive and educational measures should urgently be undertaken to protect against SHS via radical reduction of smoking.

### 1.7 Fetal programming of infant neuromotor development

Neuromotor impairment can be caused by damage to the immature brain during delivery or by medical interventions preformed after birth. However, it is more likely that deviances in the brain development originate before birth. A theory that relies on this early origin is the "Fetal Programming Hypothesis" [15]which states that fetuses adapt to limited supplies of nutrition and oxygen. These adaptations program the fetus physiology, metabolism and growth, increasing the risk or later diseases: cardiovascular and mental problems[16]

Clinical deteriorations are determined primarily by the gestational age of disease onset and placental blood flow resistance[17]. Fetal acidemia carries a greater risk of irreversible developmental delay[18], rather than hypoxemia. Delivery timing and compromise degree at birth can modify infant neurodevelopment (GRIT- Growth Restriction Intervention Trial). The proportion of essential substrates that are metabolized aerobically in the liver and their ability to drive the endocrine growth axis of the fetus is influenced by the degree of placental dysfunction. When nutritional deficiency is severe enough or has persisted for a considerable long period of time, the growth rate of all fetal measurements slows down and the sonographically estimated fetal weight eventually drops below the $10^{th}$ percentile [19]

The metabolic status and the diminishing supply of glucose force the brain and heart to metabolize lactate and ketones as their primary energy sources[20]. With increasing severity of placental dysfunction the transfer of these important nutrients also becomes impaired and their deficiency is linked independently to a range of neurodevelopmental disorders[21]. The rate of deterioration of cardiovascular parameters determines the overall speed of deterioration in early-onset FGR (Fetal Grow Restriction), often needing preterm delivery. According to this, fetuses are forced to make critical adjustments in their cerebral metabolism of essential nutrients prior to delivery. Although term FGR (Fetal Growth Restricted) does not present the same degree of clinical deterioration as early-onset disease does, abnormal brain microstructure and metabolism have been documented independently of the degree of vascular Doppler abnormalities, probably reflecting the increasing sophistication of central nervous system with accelerating synapse formation.

In summary, there is evidence that placental dysfunction is associated with delayed achievement of behavioral state organization prior to deterioration of fetal status. A recent study documented suboptimal scores for social-interactive, attention capacity, state organization and motor skills among growth-restricted neonates that had abnormal prenatal MCA (mean cerebral artery) Doppler studies[22]

To avoid bias, as to the cause of neurological impairment, in our studio, we have discarded all newborns below the 10pt of weight.

In our previous study, logistic regression analysis showed that the offspring of mothers with gestational hypothyroxinemia had an adjusted odds ratio of 3.9 (IC 95%, 1.1-14.2 p=0,036) of having an abnormal ADHD test score. But it could be argued that the age of the studied children was under the recommended age for making a diagnosis. Today, four years later, we included the 58 pregnant women of the previous study, plus 12 selected mothers who present lower values of T4 in order to increase the difference between the two study groups.

## 2. Aim

The aim of the present study is to compare the scores obtained from testing for ADHD in children whose mothers had low T4 (< 0, 79) values in third trimester of pregnancy.

## 3. Methods

ADHD test was applied to the mothers by a trained interviewer and the answers obtained from the set of questions, were expressed as a score according to the detection of the different symptoms of ADHD. The test included the tree different dimensions of the syndrome: attention, hyperactivity and impulsivity, with a total of 18 items. The diagnosis of ADHD according to DSM IV requires the existence of 6 symptoms of inattention and 6 symptoms of hyperactivity-impulsivity; although the positive diagnosis of ADHD requires a more complex psychological evaluation together with additional information obtained from the parents and school teachers during a six month period, a test a high score of which is highly suspicious of ADHD.

### 3.1 Population sample

Pregnant women who were attended during 2003-2004 in the Hospital Clinic of Barcelona, a tertiary referral hospital which covers a geographical area of mild iodine deficiency[23] participated in the study and were evaluated for thyroid function in the third trimester of gestation. Those who had history o thyroid problems as they had already received antithyroid and/or thyroid hormonal treatment were excluded of this study. The first study group was formed by only those who presented FT4 values less than 0,79 ng/dl, which is the percentile 10 of the distribution of T4 in pregnant women in our area in the third trimester of pregnancy[24] The final study group accomplishing the aforementioned criteria was composed by 40 women. The control group (n= 31 was selected in a randomized fashion by means of the SPSS program among 442 women of an original cohort who had normal FT4 values. Eventually a total of 69 women and 71 children were analyzed

Each mother gave her consent for participation in the study and permission was obtained from the Ethics Committee of the Hospital

### 3.2 Measurements

FT4 was measured by using an immunoassay (ADVIA-Centaur, Bayer) with a CV of 5.4%

In order to evaluate the syndrome, a test from the 4$^{th}$ edition of the revised manual for the diagnostic and gravity of mental illnesses was used[25]. The ADHD test was administered to the mothers by a trained interviewer which did not know if the participant was included in the study or control group.

## 4. Results

No significant differences were observed between the study group and the control group in relation to maternal age, gestational age at the time of deliver, birth weight, sex of child and Apgar scores at one and five minutes:

Children whose mothers had low levels of FT4 showed a significantly higher average score in the total score of ADHD test; also, they had higher scores in the scales of inattention and impulsivity when evaluated separately.

Unlike the previous study, the hyperactivity scale, when evaluated independently is not significant; which is consistent with the clinic, because as the child gets older the symptoms of hyperactivity decrease.

In the study group, three children were diagnosed and on treatment. One of those kids have mental retardation and laxity in the hands. The other child has serious family problems. Eight children showed high test score. Of these, one is stuttering and two have hearing loss. Two more children were being studied under suspicion of having the syndrome by a professional

Within the control group a child was diagnosed with attention deficit disorder, and in the other child, the syndrome was ruled out after psychological evaluation

| | FT4<0,79 ng/dl N=40 | FT4≥0,79 ng/dl N=31 | p |
|---|---|---|---|
| Maternal age (mean) | 32,6 | 31,4 | 0,256 |
| Weeks delivery (mean) | 39,05 | 39,8 | 0,032 |
| Birth date (mean) | 13.03.2004 | 27.08.2004 | 0,018 |
| Birth weight (mean) | 3310 | 3488 | 0,122 |
| Sex newborn (mean) | 1,58 | 1,42 | 0,199 |
| Apgar 1′ (mean) | 8,90 | 8,84 | 0,677 |
| Apgar 5′ (mean) | 9,95 | 9,87 | 0,535 |

Table 1. Characteristics of the study groups

| | FT4<0,79 ng/dl N=40 | FT4≥0,79 ng/dl N=31 | 95%Confidence Interval for mean | p |
|---|---|---|---|---|
| Inattention score; (mean; SD) | 20,55; 8,74 | 13,5 ; 6,14 | 15,5-19,5 | 0,000 |
| Hyperactivity score (mean; SD) | 14,33; 5,45 | 10,87; 5,37 | 11,4-14,2 | 0,010 |
| Impulsivity score (mean; SD) | 10,48; 4,18 | 5,97; 2,78 | 7,5 - 9,5 | 0,000 |
| Total score (mean; SD) | 45,35;16,04 | 30,39; 12,05 | 42,6-17 | 0,000 |

SD: Standard Deviation

Table 2. Score of ADHD test

## 5. Discussion

The ADHD is a neurobiological disorder with a multifactorial origin, where heritability is of 0, 8 and is of polygenic character. There are several environmental factors that could increase its frequency in which hypothyroxinaemia during pregnancy may play a role. In our study, we found an association between low maternal thyroxine levels at third trimester and a high ADHA score in the children of those mothers.

In humans, the relationship between maternal nutrition and the incidence of neural-tube defects has been actively explored. Smithells et al[26] conducted a controlled study of supplementation with multivitamins, including folic acid, to test their effect on prevention of neural-tube defects. The results demonstrated that women who received multivitamin supplementation before conception and during the first 2 months of pregnancy had significantly reduced the risk of neural-tube defects compared to unsupplemented women.

The revision of the data above, lead to the conclusion that the effect of iodine deficiency on the human fetal brain, occurred during the second (and probably) third trimester. The duration of hypothyroidism during development may crucially affect neurological disability.

One of the most important experiments of nature in regard to brain development is the neuropathological picture in endemic cretinism, in which maternal hypothyroxinemia and fetal hypothyroidism, both induced by iodine deficiency, combine during a critical period of fetal development, produce a critical degree of thyroxin deficiency severe and prolonged enough to cause irreversible damage to the ongoing program of neural development.

Fortunately, the application of different iodization programs has paid off. But both environmental pollution and depletion of croplands make that moderate iododeficiency persists in some areas of the planet. It also increases autoinmune thyroid diseases, which can cause hypothyroxinemia in situations of high demand such as pregnancy.

On the other hand, a recent survey estimates an increase of about 22% in ADHD from 2003 to the most recent survey in 2007-08. This data was obtained by the Centre for Disease Control and Prevention (CDC) that interviewed parents who had children between the ages of 4 and 17.

Researchers calculate that 5.4 million kids have been diagnosed with ADHD, which suggests that the amount of children with this disease has increased by 1 million in the last few years. Scientists don't have a clear answer about why there has been such a significant increase. Study lead author Susanna Visser of the CDC, suggests that a greater awareness of the problem in society and stepped-up screening efforts is part of the explanation.

We can no longer run the risk of letting any more children - including those with ADHD- fail at school. Their failure reflects in not only the recent increase in "dropping-out" unemployment but also reflects in a major claim of welfare benefits, more problems with authority and penal incarceration. It also privates society of the contribution these individuals could have made if their potential been developed.

In 1999 Pop[27]found significant physical and psychomotor abnormalities in children who had mothers with hypothyroidism at 12 weeks of gestational age, where as at 32 weeks no differences were found, and this is possibly due to the fact that between 16-20 weeks, the foetus begins to synthesize its own thyroid hormones and mothers with a defective thyroid function in week 32, were having an acceptable thyroxine production during the first trimester.

In our study we only have T4 values of the third trimester, However, performing a prospective study that followed up to ten years or more hipotiroxinemic mothers, could be a very interesting way to answer definitely the question if low t4 values are related with the presence of ADHD in the progeny of those mothers.

The best assessment in order to avoid this chronic disease that would need treatment for life is prevention. Measures can include: Reduction of intrauterine pollution, appropriate supplementation of the pregnant mother with vitamins and a universal screening of thyroid function in pregnant women within their reproductive age, for the correction of gestational hypothyroxinemia at the right time of pregnancy.

This will be the only way to decrease the portion of ADHD related to hipotiroxinemia.

## 6. References

[1] Hauser P, Zametkin AJ, Martínez P, Vitiello B, Matochik JA, Mixson AJ, Weintraub BD 1993. Attention Deficit and Hyperactivity Disorders in people with generalized resistance to thyroid hormones. N Engl J Med 328:997-1001

[2] Vermiglio F, Lo Presti V.P, Moleta M, Sidote M. Tortorella G. Scaffidi G, 2004 Attention Deficit and Hyperactivity Disorders in the offspring of mothers exposed to mild-moderate iodine deficiency: A possible novel iodine deficiency disorder in developed countries. J Clin End Metab 89:6054-6060

[3] Lavado R, Aus6 E, Arufe MC, Escobar del Rey, Berbel P, Morreale de Escobar G. Hypothyroxinemia induced by a low iodine diet, alters cortical cell migration in rats: an experimental model for human neurological cretinism. Eur. J Neurosci 2000; 12(Suppl 11):43(abst #121)

[4] Glioner D, de Nayer P, Bourdoux P, et al: Regulation of maternal thyroid during pregnancy. J Clin Endocrinol Metab 71: 276,1990

[5] Shapira Y, Agmon-Levin N, Shoenfeld Y. Defining and analyzing geoepidemiology and human autoimmunity. J Autoinmun 2010;34:J 168-177

[6] Bougneres P, Valleron AJ- Causes of early-onset type 1 diabetes: toward data driven environmental approaches. J Exp med 2008;205:2953-7

[7] Hewagama A. Richardson B. The genetics and epigenetics of autoimmune diseases. J Autoinmun 2009; 33:3-11

[8] De Freitas CU, Grimaldi Campos RA, Rodrigues Silva MA, Panachao Mr, de Moraes JC, Waissmann W, et al. Can living in the surroundings of a petrochemical complex be a risk factor for autoinmune thiroid disease? Environ Res 2010;110:112-7

[9] Uzma N, Salar Bm; Kumar BS, Aziz N, David MA, Reddy VD. Impact of organic solvents and environmental pollutants on the physiological function in petrol filling workers. Int j Environ Res public health 2008;5: 139-146

[10] Langer O, Kocan A, Tajtalova M, Koska J, Radikova Z, Ksinantova L, et al. Increased thyroid colume, prevalence of thyroid antibodies and impaired fasting glucose in Young aduls from organochlorine cocktail polluted: outcome o trasgenerational transmission? Chemosphere 2008; 73:1145-50

[11] Langer P, Locan A Tajtakova M, Petrik J, ChovancovaK, Drobna B et al. Fish from industrially polluted freshwater as the main source of organochlorinated pollutants and increased frequency of thyroid disorders an dysglycemia. Chemosphere 2007; 67:S379-385.

[12] Gu JY, Quian CH, Tang W, Wu SH, Su KF, Scherbaum WA, et al. Polychlorinated biphenyls affect thyroid function and induce autoimmunity in Sprague-Dawley rats. Horm Metab Res 2009; 41:471-4

[13] Effraimidis G, Tijssen JG, Wiersinga WM. Discontinuation of smoking increases thee risk for developing thyroid perixidase antibodies and/or Thyroglobulin antibodies: a prospective study. J Clin Endocrinol Metab 2009, 94: 1324-8

[14] Carrillo AE, Metsios GS, Fouris AD. Effects of secondhand smoke on thyroid function. Inflamm Allergy Drug Targets 2009; 8:359-63

[15] Barker DJ 1995 Fetal origins of coronary heart disease BMJ 311; 171-174

[16] St Clair d, Xu M, Wang P, Yu Y, Fang Y Zhan F, Zheng X, Gu N, Feng G Shan P, Hel 2005 Rates of adult schizophrenia folloeing prenatal exposure to the Chines famine of 1959-1961. JAMA 294: 557-526

[17] Oros d, gigueras F, Cruz-Martinez R, Meler E, Munmany M, Gratacos E. Longitudinal changes in uterine, umbilical and fetal cerebral doppler índices in late-onset smoall-for-gestational age fetuses. Ultrasound obstet Gynecol 2011;37:191-195

[18] Romero R, Kalache KD, kadar N. Timing the delivery of the preterm severely growth-restricted fetus:venous doppler, cardiotocography or the biopysical profile? Ultrasound obstet Gynecol 2002; 19:118-121

[19] Baschat AA. Fetal growth restriction-From observation to intervention. J Perinat Med 2010; 38: 239-246

[20] Vannucci RC, Vannucci SJ. Glucose metabolism in the developing brain Semin Perinatol 2000; 24: 107-115

[21] Ward PE. Potential diagnostic aids for abnormal fatty acid metaboliism in a range of neurodevelopmental disorders. Prostaglandins lukot Essent Fatty Acids 2000; 63: 65-68

[22] Cruz-Martinez R, Figueras F, oros D, Padilla N, Meler E, Hernandez-Andrade E, Gratacos E. Cerebral blood perfusión and neurobehavioral performance in full-term small-for-gestational-age fetuses. Am j obstet Gynecol 2009; 201: 474.e1-7

[23] Vila L, Castell C, Wengrovicz S, de lara N and Casamitjana R. Urinary iodide assessment of the adult population in Catalonia (2006) Med Clin (Bar) 2006;127:730-3-vol127 num 19

[24] Vila L, Serra-Prat M, De Castro A, palomera E, Casamitjana R, Muñoz M, de Castro B and Puig-Domingo M (2006) references values of FT4 and TSH in pregnancy: Is FT4 the best parameter? Endocrinol Nutr 53, 138

[25] Diagnostic and statistical manual of Mental Disorders. (DSM-IV) American Psychiatric Association 1994

[26] Smithells RW,Shepherd S, Schorach CJ. Et al. Vitamin supplementation and neural tube defects. Lancet 1981; 2: 1424-5

[27] Pop V, Kuipens Jl, Van Baar AL, Verkerk G, Van Son M, de Vijlder JJ, Vulsma T, Wiersing WM, Drexhage HA, Huib L and Vader Hl 1999. Low normal maternal free T4 concentrations during early pregnancy are associated with impaired psychomotor development in infancy. Clin Endocrinol (Oxf) 50:149-155.

# 6

# Cutting Corners: Neuropsychological Research into the Energetics of ADHD

Jens Egeland
*Vestfold Mental Health Care Trust, Institute of Psychology, University of Oslo, Norway*

## 1. Introduction

People with ADHD are often subject to normative judgments by others for not doing their best, or being lazy. At school, the teachers' urge that they should put more effort into school work, may often be phrased as a question of *will* rather than lacking *ability*. Figuratively they could be said to cut corners, in the sense that they often perform tasks the easiest way. Even though they may know the long term advantages of hard work or of performing dull tasks, they may nonetheless follow the desires of the moment. This deficit in volition or self control has been construed as an aspect of behavioural inhibition or executive functioning (EF) (Barkley, 1997) considered as a core deficit in ADHD. Executive functions exert a top top-down mediating role on other cognitive processes, such as attention, learning and motor function. An ongoing discussion is whether impaired top-down processing is sufficient to explain ADHD, or sufficient to explain the heterogeneity of ADHD symptomatology. For some time, most neurocognitive research has focused on EF functions, clearly showing that subjects with ADHD are deficient in such tasks. However, there are four problems with considering this EF impairment as the cause of disordered behaviour.

### 1.1 Problems with EF impairment as a core deficit of ADHD

First: Although the research following this line has shown unrefutable results, effect sizes are at best moderate and only account for a part of the diagnosis related variance (Banaschewski et al., 2004; Toplak et al., 2009).

The second problem is related to specificity: other neurocognitive disorders also display dysfunctions of EF while clearly displaying other behavioural problems (Salimpoor & Desrocher, 2006). Zelazo and Müller (2002) suggest that EF deficits are a common outcome of many different perturbations of the epigenetic process, rather than a cause. They suggest that deficits in planning and inhibition, often related to a dysfunctional dorsolateral prefrontal cortex (DL-PFC), may be secondary to a developmental older dysfunction in orbitofrontal cortex (OFC) in autism, while dysfunctions in regulations of emotions and social relations in ADHD often associated with OFC, is secondary to their dysfunctions in DL-PFC. The main point of interest to this chapter, is that EF dysfunction is considered a symptom rather than a mechanism. These symptoms may be caused by a mixture of bottom- up and top-down processes or both types of processes can account for the same result.

The third problem related to the primacy of inhibition or EF dysfunction, is related to ADHD subgroups. Although described by parents and teachers as clearly different, many studies have failed to find EF differences between subjects with the combined and the inattentive subtype of ADHD (Geurts et al., 2005; Nigg et al, 2002), suggesting that other mechanisms may underlie this division.

The fourth problem is related to variability in reward contingencies. External immediate reward has been found to increase performance quality or even normalize otherwise deficient functioning among subjects with ADHD (Liddle et al. 2011; Luman et al, 2008) even though external structure and need for internal structuring i.e. EF demand, remains the same.

### 1.2 Bottom-up mechanisms in ADHD

That ADHD can be caused by both bottom-up and top-down mechanisms are the central axiom of the Dual Pathway Model (Sonuga-Barke, 2002). Similarly to Barkley, it posits inhibitory dysregulation as one pathway leading to ADHD symptoms. The other pathway has to do with motivational style and is associated with alterations in reward mechanisms. According to this view, inattention, overactive and impulsive behaviour can be functional expressions of delay aversion. When faced with a choice between immediacy and delay, ADHD children will choose the former (Sonuga-Barke et al, 1992). When no choice is available, they will reduce the perception of time by engaging in task irrelevant or incompatible (hyperactive) behaviour or being inattentive. In this model the cognitive deficits such as impaired planning and working memory, are seen as secondary to the bottom-up effects of delay aversion for some subjects with ADHD. For other children with ADHD poor inhibitory control underlie much of the same symptoms. Sonuga-Barke (2002) considers the motivational and the regulation dysfunction as independent, thus the dual pathway. In a study of cognitive impairments in probands with ADHD and their relatives, Kuntsi et al. (2010) found indications of the existence of two familial distinctive patterns. Another study indicated the independence of three pathways, namely temporal processing as well as inhibitory control and delay aversion (Sonuga-Barke et al., 2010).

Recent research on default mode network (DMN) in ADHD gives support to the emphasis on bottom-up mechanisms in ADHD. DMN is a distributed brain system comprising medial prefrontal cortex and medial and lateral parietal regions that are anticorrelated with attentional networks. While attentional networks are activated by goal-directed behavior (Liddle at al., 2011), activity in the DMN is related to mindwandering and resting state. When performing a task, the DMN must be deactivated. Both fMRI and EEG studies have found impaired attenuation of the DMN from rest to mentally demanding tasks among subjects with ADHD (Helps et al., 2010; Liddle et al., 2011;Peterson et al., 2009).

The emphasis on heterogeneity of mechanisms giving rise to the symptoms of ADHD is less clear in the Cognitive Effort Model (CEM) compared to the Dual Pathway Model. Similarly to the Delay Aversion pathway, it emphasizes the impact of bottom up motivational processes (Sergeant et al., 1999). Sergeant et al. (2003) considers delay aversion and aberrant reward systems as one expression of deficient energetics. Sonuga-Barke et al.(2010) agrees that the two theories overlap, but that it is possible to deduct different predictions that can be tested in future research.

Sergeant et al. (2003) present the CEM model as the most comprehensive model, incorporating also the top-down processes emphasized by Barkley (1997) and by Pennington & Ozonoff (1996), and extending the view of energetics in larger detail. The model divides the energetic of ADHD into different resource pools, i.e. effort, activation and arousal (Sergeant, 2005).

This chapter will present clinical research on the energetics of ADHD. First, the neurobiological basis and the psychopharmacological evidence will be briefly outlined, before proceeding to our own clinical neuropsychological findings. The energetic resource pools are in this connection considered latent variables difficult to operationalize in ways that differentiate them from aspects of executive function. Based on our own previous research, we nevertheless think it is possible to deduct measures from neuropsychological examinations that can serve as approximate operationalizations, making it possible to extract their contributions to some of the problems affecting subjects with ADHD.

## 2. Energetic resources

The first level of CEM comprises different stages in information processing: Information must be detected and encoded, be subjected to some type of processing and responses must be organized. The three energetic resources modulate these processes. *Effort* is required whenever the current state of the organism does not match that required to perform the task (Sergeant, 2005). Effort encompasses factors such as motivation and response to contingencies. The claim that effort seems to be impaired in ADHD is based on findings of variability of performance. Under specific circumstances that increase intrinsic motivation or under high extrinsic reinforcement, subjects with ADHD increase their performance more than controls (Luman et al., 2007). If reinforcement and motivation contingencies rather than task complexity are decisive, the typical impaired performance of ADHD subjects on effortful executive function tasks, would better be described as impairment in effort than in executive function per se. The research on the DMN is relevant as a possible physiological mechanism, indicating that subjects with ADHD fail to down-regulate the resting state and up-regulate the necessary effort related activation necessary for effective performance.

The other energetic pool, *arousal* is related to the alerting effect of sensory activity. Signal intensity and novelty increases arousal. By the same token, decreased intensity of stimulation leads to falling arousal. Arousal is the time-locked phasic physiological response to input and is regulated from the frontolimbic forebrain and the by basal ganglia. The primary neurotransmitters are noradrenaline og serotonin (van der Meere, 2002). The third energetic resource pool, *activation,* is related with the tonic or long lasting physiological readiness to respond and is affected by task variables such as preparation, alertness, time of day and time on task (Sergeant, 2005). While stimulation increases arousal, activity increases activation. The primary neurotransmitters are dopamine and acetylcholine.

### 2.1 Neuropsychological operationalizations of energetic resources

An operationalization of effort allocation in behavioral or neuropsychological terms needs to refer to a task in which the subject can *choose* how much effort to put into the task, in the sense that the environmental constraints on task performance must be minimal. If it is too easy, we would not see any potential effect of a failure of effort allocation. If it can only be solved by putting in high effort, we will not know whether a failure to perform successfully was due to task complexity or the proposed deficit in state regulation or effort allocation.

In order to interpret an ADHD related deficit in terms of effort allocation, it is also necessary to avoid testing effort with measures of executive function or attention. Such tasks are generally effortful, but impaired performance is difficult to interpret as impairment in motivation and not in executive function per se. In a recent study applying Evoked Response Potentials (ERP) and Skin Conductance Level (SCL) Johnstone et al. (2010) tried to disentangle the effect of effort and EF. As in many other studies of neuropsychological function in ADHD, they contrasted the behavior and ERP/SCL of subjects with ADHD to controls on attention to congruent and incongruent stimuli. Withholding response to incongruent stimuli would require interference control and is considered effortful. However, intending to make a critical test of the predictions from CEM and the interference control model of Barkley, they reasoned that degrading stimuli both in congruent and incongruent stimuli presentations would make the task more effortful while not increasing demand for interference control. They found an effect of stimuli degradation on skin conductance level, indicating that degrading stimuli increased arousal level. Also ERP findings were interpreted as suggesting impaired resource allocation (i.e. effort) rather than interference control per se. However, contrary to what we found, they did not find any significant group effects on the performance level.

The third condition that must be met in designing a performance measure of effort, is that it must not be a test of general ability in disguise. In our own research on learning and memory in ADHD, to which I soon will turn, we think that we constructed measures of effort that satisfy these three criteria (Egeland et al., 2010).

Research on the effects of event-rates on continuous performance tests (CPTs) has shown that subjects with ADHD tend to be impaired with large inter-stimulus intervals (van der Meere, 2002). This has been interpreted as an effect of underactivation (Sonuga-Barke et al., 2010), but might as well be related to underarousal. It is however, difficult to discern the effect of stimulus intensity from activation effects as long as measures demanding motor responses are the dependent variable. Behaviorally we cannot register reactions to stimuli the subject have to attend to, but not respond to. The finding of Drechsler et al. (2005) that ADHD subjects responded slower to targets not preceded by a warning signal, indicates an effect of arousal rather than activation. In the same vein, Benikos and Johnstone (2009) found that the length of the interstimulus-interval not only had effect on performance to targets, but also to ERP responses to a warning signal *preceding* the target that was responded to.

In the author's opinion previous interpretations of CPT performance have not sufficiently differentiated between the input or output aspects of the information processing, i.e. between stimulus and response intensity. In older CPTs, stimulus intensity have either been fixed or not analyzed in standard scoring systems. In these so called "low stimulus to noise" tests the subjects are to watch stimuli appearing on the computer screen and respond to about 10 or 20 % of all stimuli. The Conners' Continuous Performance test (CCPT-2: Conners, 2002), however, has a high stimulus to noise ratio. Subjects are supposed to respond to *all* letters appearing on the screen, *except* to X's, which amount to 10 percent of the exposures. At the same time, the inter-stimulus intervals (ISI) vary between 1, 2 and 4 seconds. This makes it possible to differentiate between the phasic effects of high or low stimulus intensity while at the same time assess the long term activation effects, since the subject has to respond more actively than in the former generation of CPTs.

In their seminal paper differentiating different aspects of attention, Mirsky, Anthony, Duncan, Ahearn & Sheppard (1991) factor-analyzed results from a series of attention tests, including a low signal to noise CPT. They identified four sub-functions of attention. The scores from CPT loaded on what was termed a vigilance factor. Vigilance was defined as the attentional capacity to remain alert also when less stimulated, i.e. similarly to loosing arousal, but was considered synonymous to sustained attention. As was the custom of the time, the sum of CPT omissions and commissions was analyzed with no analysis of time on task changes or changes after different interstimulus-intervals. Since the Conners's CPT offers a lot of subtle measures not available in the older CPTs we hypothesized that it would discern different patterns of attention deficit among different patient populations. Thus, we sat out to perform the new factor analysis to which we now will turn.

### 2.1.1 The factor structure of Conners' CPT-II.

CPTs are widely used in neuropsychological assessment of subjects with ADHD (Wasserstein, 2005) as well as in schizophrenia research. Many studies have shown that children (Root & Resnick, 2003) and adults (Hervey et al, 2004) with ADHD are impaired compared to normal controls on this type of test, but also other clinical groups show impairments. In fact, the lack of differences between clinical groups, represent a problem when using the test as part of an ADHD assessment. Studies have failed to find differences between clinical groups such as ADHD and reading disorder (McGee et al, 2000), ADHD and schizophrenia (Øie & Rund, 1999) and ADHD and internalizing disorders (Solanto et al, 2004). These studies have, however, analyzed overall measures of attention from the CPT, such as d', i.e. signal detection that is derived from both omissions and commissions and have not taken into account whether the errors are performed initially or late in the test, or subsequent to short or long stimulus-intervals. The reason for not being able to differentiate between clinical groups may then not be a lack of reliability, but may rather represent an accurate description that all these groups also suffer from a deficit in attention. Although they fail in attention for different reasons. While hyperactivity and impulsivity are considered to mediate the attention deficit of subjects with ADHD (Epstein et al, 2003), subjects with schizophrenia may have an impairment initially focusing attention, but may profit from exercise (Egeland et al., 2003, Egeland et al., 2007). Finally, lack of effort or fatigue may underlie the attention deficit in depression (Egeland et al., 2003). It is reasonable that the different mechanisms underlying the attention deficit will also be reflected in different patterns of CPT performance. The factor-analysis described below (Egeland & Kowalik-Gran, 2010a) was performed on CCPT-II protocols from a mixed clinical sample of 376 adolescent and adult participants with either ADHD-C, ADHD-I, affective disorders, schizophrenia spectrum, mild mental retardation or mild neurocognitive disorder, nonverbal learning disorder, learning disorder, different mild psychological disorders and subjects using analgetics. A normal group was included as well. In a follow up study to validate the factors (Egeland & Kowalik-Gran, 2010b), hypotheses were formulated as to the how subjects with ADHD, schizophrenia, affective disorders, language disorders and brain injury should perform in order to consider them different sub-processes. As part of the validation procedure, correlations with other tests of attention were also computed.

The result of the factor analysis is presented in Table 1. The first factor was coined *focused attention* since it was an overall measure of being able to focus on the task, namely to respond quickly whenever target stimuli appeared. Previously, studies had shown problems with focusing attention among all the analyzed patient groups, and thus it was expected that they could not be differentiated on this factor. It was also expected that this factor should correlate with other tests of controlled attention such as the Stroop Color-Word interference Test, the Trail Making Test and the Paced Auditory Serial Addition Test, but not with Digit span or Knox Cubes measuring more automatic attention span. These hypotheses were all confirmed.

| | I Focus | II H/I | III Sustain | IV Vigilance | V Change in control |
|---|---|---|---|---|---|
| Variability[1] | **.870** | .086 | .133 | .194 | -.010 |
| Hit RT SE[2] | **.843** | .265 | .145 | .334 | .056 |
| Perseverations[3] | **.768** | .612 | .019 | -.093 | -.033 |
| Omissions[4] | **.747** | .255 | .224 | .058 | .072 |
| Commissions[5] | .335 | **-.804** | .048 | -.022 | -.123 |
| Hit RT[6] | .363 | **.760** | .049 | .251 | .185 |
| Response style[7] | .206 | **.688** | .089 | -.012 | -.087 |
| Block change SE[8] | .162 | -.064 | **.842** | .128 | .020 |
| Block change[9] | .068 | .047 | **.731** | .073 | **.469** |
| Δ omissions[10] | -.137 | -.131 | **-.707** | .042 | .245 |
| Hit RT ISI[11] | .097 | .021 | .014 | **.847** | .163 |
| Hit RT ISI SE[12] | .135 | .103 | .099 | **.842** | -.042 |
| Δ commissions[13] | .197 | .086 | -.022 | .088 | **.904** |
| Eigenvalue | 3.94 | 1.93 | 1.49 | 1.31 | 0.99 |
| % variance explained | 22.92 | 15.13 | 14.28 | 12.94 | 9.17 |

[1] Variability of standard errors i.e. a measure of within respondent change in consistency of reaction time. [2]Hit reaction time standard error (i.e. consistency of response time). [3]Perseverations: responses without preceding stimuli. [4]Omissions: missed targets, [5]Commissions: responses to non-target stimuli . [6]Hit reaction time. [7]Response style (β): cautious response style aimed at minimization of commission errors, or impulsive style minimizing omission errors. [8]Hit reaction time standard error over time-blocks (change in consistency as the test progresses) [9]Block Change: the slope of change in reaction time over six time blocks. [10]Δ omissions: Changes in omissions over time (numbers of omissions in last third of test subtracted from first third).[11] Hit reaction time ISI: decrease in reaction time with longer interstimulus-interval.[12]Hit RT ISI SE: Standard Error of Hit RT ISI, i.e. whether reaction time becomes more variable with longer interstimulus intervals). [13]Δ commissions: Changes in commissions over time.

Table 1. **Factor structure of Conner's CPT** (from Egeland & Kovalic-Gran, 2010a)

The next factor was termed *hyperactivity-impulsivity* (H/I) since it received loadings from commissions and reaction time. To validate such a term for this factor, we would expect that only the ADHD-C group would perform impaired. None of the other groups were expected to have a H/I problem. To qualify as a measure of H/I scores it should also not correlate with any of the other tests of attention. Also these hypotheses were confirmed.

The remaining three factors measured attributes of attention that had to do with changes over time or stimulus contingency. The two block-change-measures, as well as Δ omissions (the difference between number of omissions in the first and the last third of the test), were named *sustained attention*. The ADHD-I group scored below all other groups on this measure. Differential validity came from the finding that the factor did not correlate with any other attention test intended to measure other aspects than sustaining attention per se. The two ISI measures (changes in reaction time or increased variability of reaction time as a function of increased interstimulusinterval) loaded on the fourth factor named *vigilance*. Contrary to expectations, there was only a non-significant tendency for the ADHD-C subjects to score below the normal group. The fifth factor received a high loading from Δ commissions, and a mediocre loading from block change. The factor seemed then to measure whether subjects became more impulsive as a function of time on task, and it was thus termed *change in control*. Both ADHD- groups scored below the normal group as well as the schizophrenia spectrum group and subjects using analgetics. That also subjects with ADHD-I lost control was contrary to expectations, but as a change measure, it could be due to different initial level of commissions. Interestingly, since previous studies had shown difficulties to differentiate between subjects with LD and ADHD when using overall measures, this process measure showed that LD subjects gained control, whereas ADHD-I subjects lost control. This is reasonable if the LD subjects had commission errors due to difficulties with letter differentiation, whereas commission errors reflect impulsivity among ADHD subjects.

Overall, the study showed that different mechanisms mediate the attention deficit in different groups, and that H/I is specific to ADHD-C, while impaired sustained attention is specific to ADHD-I. It also differentiated sustained attention from vigilance, thus possibly giving the clinician a tool to distinguish between arousal and activation.

### 2.1.2 Differentiating arousal and activation

This was examined further in a study of Conners' CPT performance comparing children and adolescents with ADHD-C and ADHD-I and healthy controls (Egeland et al., 2009). Sixty-five healthy controls and 67 subjects with ADHD between nine and 16 years of age participated in the study. The ADHD-I group performed below control children on Hit Reaction Time Block Change, considered to measure sustained attention. The ADHD-C group scored below controls on Hit Reaction Time Inter-Stimulus-Interval, considered to measure vigilance.

As illustrated in Figure 1, comparison of the two clinical groups showed a test by group interaction, with ADHD-I subjects performing below ADHD-C subjects with regard to sustained attention and above ADHD-C subjects with regard to vigilance. Sustained attention on the CCPT correlated specifically with parent and teacher ratings of inattention, but not with ratings of H/I, while vigilance correlated with all symptom ratings. Although correspondence between general findings of attention deficit in neuropsychological laboratory tests and daily life ratings tend to be significant but mediocre (Toplak et al, 2009), it is seldom that specific measures intended to measure underlying mechanisms mediating the deficit, also correlate with daily life ratings. In this case, the measure possibly reflecting insufficient activation correlated only with inattention scores and not with hyperactivity. However, the measure intended to quantify low arousal correlated also with H/I. This could be an indication that H/I is a way for the ADHD-C subjects to compensate for low arousal, but that prolonged activation leads to the fatigue more typical of the ADHD-I subjects. Contrary to the expectation from Barkley's interference control model, numbers of commissions neither differentiated between the ADHD subgroups nor between ADHD and healthy controls in this

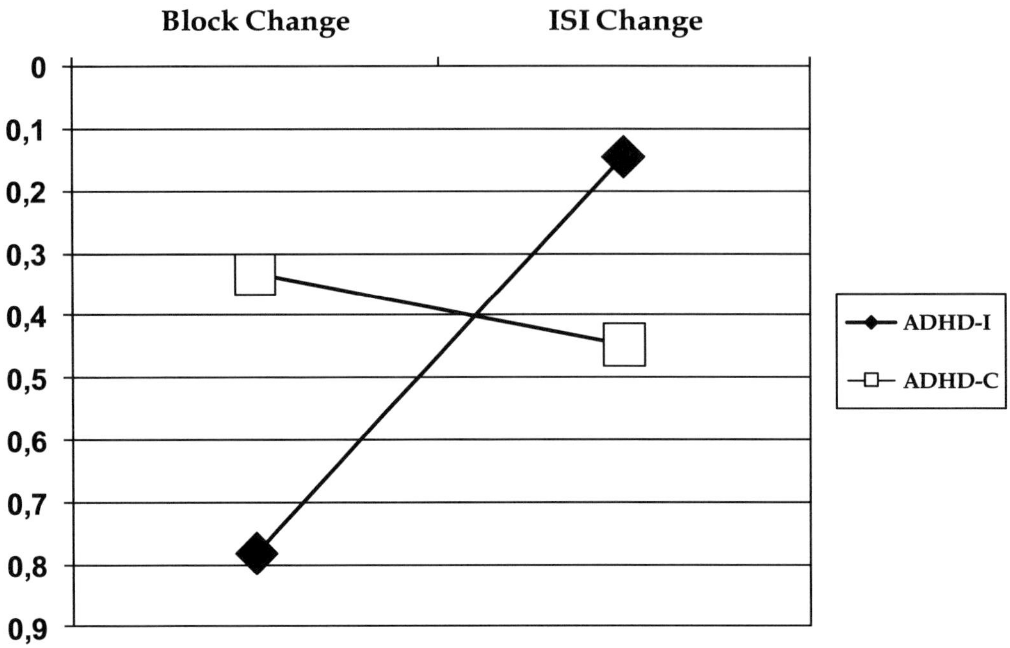


Effect size in z-scores derived from age matched healthy controls

Fig. 1. **Double dissociation between presumable arousal-mediated vigilance (ISI Change) and sustained attention in ADHD-C and ADHD-I** (from Egeland et al., 2009)

study. However, Barkley et al. (2001) also used the CCPT-II and found no group difference between ADHD-C subjects and healthy controls with regard to commission errors. They commented that their results contradicted previous findings and could have to do with the type of CPT applied. That the difference between low and high signal to noise CPTs alter what was previously considered a main finding regarding ADHD performance on CPTs testifies to the moderate and variable effect size of this EF measure (Banaschewski et al., 2004; Toplak et al., 2009).

### 2.1.3 The impact of effort

The next study that will be discussed here concerns the effort construct (Egeland et al., 2010). As mentioned previously, to differentiate the bottom-up process of allocating effort to difficult tasks, from the top-down process of controlling attention, one has to construct effort measures that are not at the same time measures of executive function. Our point of departure was the observation that subjects with ADHD underperform at school (Ek et al, 2011). In formal tests of memory, they tend to be impaired in free recall, but often not impaired to the same degree in recognition memory. How could this be explained? If they are generally inattentive, how could they then be able to recognize items that they not had attended to in the first place? Could it be that encoding of new information was insufficient due to insufficient effort rather than inattention? All subjects were tested with a verbal list learning test. Four measures were computed that were considered to demand effort, without placing an equivalent load on executive functions (i.e. not demanding flexibility, inhibition, willful focusing on some information and ignoring other). The four measures were:

*Semantic organization of recall*: This was a measure of whether the subjects organized the haphazardly presented learning list in a thematically organized way when reporting back what they remembered. Such organization requires elaborative encoding considered to be specifically effortful (Hasher & Zacks,1979).

*Buildup of proactive interference* (PI): PI refers to a normal process in which previous learning interferes with new learning. It is the price one pays for deep level effortful encoding.

*Degree of retroactive interference* (RI): RI refers to the phenomenon that new learning interferes with effective retrieval of previous learning. While lack of effective learning would *prevent* the build-up of PI, the same low-effort learning would be expected to *increase* the potential for retroactive interference. While PI depends on the original learning material still being remembered, RI is in fact forgetting due to new learning overriding the effect of old learning.

*Overriding the primacy and recency effect*: When presented with a learning list, the easiest items to remember are the first and last ones. They are often remembered "for free" or automatically while remembering the middle items demand some kind of organization of the stimuli, again considered to be an effortful process.

The study showed that the ADHD-C subjects differed from the healthy controls on all four effort indices, while the ADHD-I group differed on three of them, but showed a normal proactive interference. The effects were significant even when controlling for IQ that otherwise is related to effective learning strategies. Most effects were significant also when analyzing only ADHD subjects with no comorbid conduct or oppositional defiant disorder (CD & ODD). As in most other studies, the ADHD groups were impaired in free delayed recall, which of course is most important for school learning. Regression analyses of the explanatory power of the effort measures on such delayed recall, showed that they accounted for 39 % of the variance in the ADHD-I/control-analysis, and 35 % in the analysis of the ADHD-C and controls. When IQ was entered into the regression analysis after the four effort indices, the total variance explained rose to 45 % in the ADHD-I group. When entering diagnosis after the effort-indices, the total explained variance increased only marginally. In fact diagnosis did not contribute significantly beyond the effect of effort. Such statistical analyses do not disclose causative effects, but they show that impaired effort could be a sufficient explanation of impaired learning among subjects with ADHD.

### 2.1.4 Effort and arousal effects on motor performance

In the last study that will be discussed here, we tried to integrate both effort and arousal variables as possibly underlying motor impairment in ADHD (Egeland et al., 2011). Impaired motor function was one of the criteria for diagnosing Minimal Brain Dysfunction (MBD) which was a precursor of the present ADHD-diagnosis. Although deficits in writing, drawing and fine motor function still is considered typical of ADHD, the prevailing view today is that such impairments are merely secondary to the core ADHD symptoms, be that impulsivity/hyperactivity or energetics. ADHD subjects cut corners when drawing squares not because they lack motor skills, but because of a central processing deficit, i.e. a deficit that transcends the motor domain.

The same subjects that participated in the effort-study, also took part in this study. The subjects were tested with the Visual Motor Integration test (VMI: Beery, 1997) in which they were required to draw 27 drawings as neatly as possible. Difficulties with this, which is typically found (Geurts et al, 2005), could be due to lacking motor function or to not allocating sufficient effort to the task. The reason for this, was that they themselves could choose how long time they would spend on the test and decide on their own level of accuracy within the limits of their capabilities. If motor problems mediated impaired performance, VMI performance should correlate primarily with other tests of motor function where energetics is not probable to influence results. Fingertapping speed and dexterity measured in a pegboard test were chosen as such tasks. If energetics as an example of a central processing deficit influenced VMI performance, then performance on this test should correlate with effort measures and arousal even when these measures are derived from a completely different behavioral domain. A summed effort measure was computed from the four separate effort measures from the previously cited memory function study (Egeland et al., 2010) and the interstimulus-interval effect from Conners' CPT study was used as the arousal-measure.

The results of partial correlations controlling for age and IQ showed that the simple motor tests did not correlate with any of the energetic-measures, indicating that they represented different sources of variance. Despite this, both energetics-measures correlated significantly with VMI performance in the ADHD-C group as did the simple motor tests. In the ADHD-I group the correlation between energetics and VMI performance was close to zero. This was interpreted to mean that when subjects with ADHD-C are more impaired in a complex fine motor test such as VMI, this is due to an impairment both in simple fine motor control and energetics, while only impaired fine motor control contributed to the impairment in ADHD-I.

### 2.1.5 Critical appraisal of the performance based research on energetics

The studies presented above show that attention over time can fail for different reasons. This is not a trivial statement, as impaired attention is reported in several neuropsychiatric conditions, as well as in conditions traditionally considered to be mostly psychogenic in nature. If we want to understand ADHD, a prerequisite is a thorough understanding of the specific mechanisms mediating the attention deficit of that disorder. Whereas persons with different conditions can be impaired in focusing attention on a descriptive level, they nevertheless differ as to whether the attention deficit is characterized by hyperactivity-impulsivity, by presumably reduced arousal under low-intense external stimulation, or reduced ability to sustain activation, as found in ADHD.

An important challenge is to bridge the gap between experimental research and clinical practice. The research reported here uses standard neuropsychological tests commercially available to the clinician, while also supporting previous findings using other methods that have indicated a role for energetics. Extending the research beyond the core symptom of attention deficit, the presented research on the effect of impaired effort allocation gives a possible explanation of the learning and motor impairments accompanying ADHD. Again the findings are derived from clinical tests but nicely fit the notion of impaired attenuation of default mode processing in ADHD.

Although the research described here suggests a role for energetics, does it represent a critical test between the interference model of Barkley and the Cognitive Effort Model?

The finding of an interstimulus effect and the lack of a between group difference in commissions in the Egeland et al study (2009), at least makes the arousal interpretation more parsimonious. However, the effect sizes indicating low arousal in ADHD-C and loss of activation in ADHD-I are small, and can by far be called a defining characteristic. Heterogeneity is typical of all neurocognitive research on ADHD. Examining the frequency of individuals with clinically significant attention deficits in ADHD and schizophrenia Egeland (2010) found that only a minority of adult subjects with ADHD-I were impaired in sustaining attention. The reason for the heterogeneity may be problems with measuring a phenomenon that nevertheless is the underlying mechanism, i.e. the problem of sensitivity. In the Egeland (2010) study, however, the difference between different cut-off levels for impairment was small. Applying a cut-off level for mild impairment showed that somewhere between 35 % and 45 % percent, depending on what measure of sustained attention was used, were impaired in sustained attention. However, 30% to 35 % were impaired also when using a strict level for severe impairment. Although it is generally found that neuropsychological tests are less sensitive than rating scale information regarding behaviour related to the same phenomenon (Toplak et al., 2009), the relatively small impact of changing from the severe to the mild impairment cut-off level indicates that differences in sensibility at best explains only part of the heterogeneity. Another possibility is that there are *several* underlying mechanisms mediating typical ADHD behaviour. These mechanisms may only partly be related to the present subdivision into combined, inattentive and hyperactive-impulsive subtypes. Sonuga-Barke et al. (2010) found that a substantial proportion of ADHD subjects were only impaired in one of three behavioural domains and that their cognitive profiles correlated with that of unaffected siblings. Inhibition was the least common impairment pattern. This research points to the validity of sub-classifying, but challenges the present division.

It may also be that the phenomenon we examine is merely an often occurring symptom more than a measure of the underlying deficit. This is what we claim is the case regarding the findings related to EF and inhibition specifically. An additional source of methodological noise is related to making the diagnosis. Presently, the diagnosis is set by collecting information about function in daily life, indicating attention deficit and/or hyperactivity/impulsivity evident in at least two areas in life, for instance school and at home. It is clear that environmental conditions such as where survey information is collected (Rescorla et al., 2007) and class size (Havey et al., 2005) influence the informants interpretations of children's behaviour. This may contribute to over-inclusion of subjects with ADHD, that again will confound studies looking for underlying mechanisms mediating functional deficits.

Turning back to the presented studies, it is fair to state that although the finding of an interstimulus-effect in ADHD must now be considered established knowledge, the differentiation between the two subgroups of ADHD with regard to arousal, are new. Thus, this must be replicated preferably with imaging or brain activation-measures that could corroborate the findings from the performance based measures, and give validity to the interpretations of them as measures of arousal and activation respectively. The specificity of the arousal and activation measures must also be examined, i.e. whether the same findings characterize other neuropsychiatric conditions that otherwise display executive dysfunction, such as Tourette syndrome and autism.

While the arousal and activation effects were small, the finding of en effort effect on learning, was large. A critical question, however, is whether the operationalizations of effort could be interpreted as executive dysfunction as well. Although impaired executive function cannot we equated to frontal lobe dysfunction, the two are clearly related to the extent that frontal damage are often used as a metaphor for executive dysfunction (Pennington & Ozonoff, 1996). Imaging studies localize proactive interference to the left inferior frontal gyrus (Feredoes et al., 2006), while lesion studies generally find that subjects with frontal brain damage display a *larger* PI effect than controls. Based on a presumption of executive dysfunction one could therefore expect that subjects with ADHD should display an *increased* PI. Instead the *lack* of the normal PI effect indicates low effort processing.

The possible differentiated expectancies from Barkley's interference control model and CEM regarding the other three measures of effort, namely retrograde interference, middle list responses and semantic organization are less clear. Searching the research literature shows no linking of these organizational phenomena to interference control. However, that may reflect that researchers interested in interference control use tests designs directly involving conflict resolution, and represent no direct evidence that executive functions do not play a role in choice of learning strategies. Imaging studies show frontal activation during deep level encoding, so if one adhere to the frontal metaphor for executive dysfunction (Pennington & Ozonoff, 1996) that will indicate a role for EF, although perhaps not for interference control as a more limited EF sub-process, in such organizational processes. However, localization of mental processes is giving way for network models of cognitive function. A large metastudy of fMRI studies of memory (Kim, 2011) showed differential activation patterns between items subsequently failed and items successfully encoded. A distributed network of five brain areas constituted the activation pattern associated with successful encoding, whereas failure of encoding was related to activity in the Default Mode Network. This indicates that inability to change into the effortful task related network underlie failure of encoding.

The hypothesis that insufficient effort constitutes a behavioural correlate to impaired ability to shift from Default Mode processing to task-specific processing, must be tested in studies that apply both behavioural measures (i.e. neuropsychological tests) and imaging techniques.

Future research must also study the relations between the dopaminergic thalamo-cortico-striatal dopaminergic reward system and effort. A dysfunctional reward system in ADHD leads to weaker conditioning, faster extinction of behaviour, and a weaker influence of reinforcers on behaviour in the sense that they are controlled by immediate rather than distal reinforcers (Stark et al., 2011). The dysfunctional dopamine system can be the neurochemical mechanism underlying impaired energetics. A study by Søderlund et al. (2010) showed that white noise normalized memory performance in inattentive school children and worsened performance among attentive school children. The authors speculate that simultaneous noise in inattentive children with low tonic dopamine level, increase stimulus dependent phasic response to stimulation, while a high tonic level suppresses the phasic release. Although not designed as a critical test comparing the interference and the energetic model, one could have expected that having to suppress background task-irrelevant noise should instead lead to distraction.

## 3. Conclusion

The chapter outlines the energetics of ADHD posited by the Cognitive Energetic Model (CEM). Arousal, activation and effort are considered bottom-up processes mediating attention, response and processing capacity among subjects with ADHD. The author's neuropsychological research tries to go beyond global measures of attention, learning and motor function, and to operationalize clinical available measures of the energetics of ADHD. The presented research shows subgroup differences between ADHD combined and inattentive subtypes, indicating a role for effort allocation in both. Impaired arousal may be most typical of the combined subtype, while deficient sustained attention is more typical of the inattentive subtype. Impaired effort allocation may explain impaired memory in both subtypes, whereas impaired motor function may be secondary to impaired energetics only in the combined subtype. Generally, effect sizes are small and heterogeneity large in clinical research on ADHD. This may be due to methodological problems such as inadequate sensitivity of measures or incorrect threshold levels for diagnosis. Also more substantial reasons such as measuring the wrong construct or genuine heterogeneity with regard to what is the core deficit in the disorder, may cause variability. Integrating neuropsychological methodology with research using both time and space distributed brain imaging techniques on the DMN and dopaminergic reward systems will probably lead to a better understanding of the role of bottom-up processes in ADHD, and to what extent they represent additional or alternative explanations of symptomatic behaviour.

## 4. Acknowledgment

The author would like to acknowledge the contributions of my colleagues from Vestfold Mental Health Care trust, taking part in the presented research. These are Iwona Kowalik-Gran, Susanne Nordby Johansen and Torill Ueland.

## 5. References

Banaschewski, T.; Brandeis, D.; Heinrich, H.; Albrecht, B.; Brunner, E. & Rothenberger, A. (2004). Questioning inhibitory control as the specific deficit of ADHD - evidence from brain electrical activity. *Journal of Neural Transmission, 111,* 841-864.

Barkley, R.A. (1997). *ADHD and the nature of self-control.* New York: The Guildford Press.

Beery, K. (1997) the Beery-Buktenica Developmental Test of Visual Motor Integration: VMI with Supplemental Tests of Visual Perception and Motor Coordination: Administration, Scoring and Teaching Manual. Parsipany, N.J.: Modern Curriculum Press, 1997.

Benikos, N. & Johnstone, S.J. (2009). Arousal-state modulation in children with AD/HD. *Clinical Neurophysiology, 120,* 30-40.

Drechsler, R.; Brandeis, D.; Földényi, M.; Imhof, K. & Steinhausen, H-C. (2005). The course of neuropsychological functions in children with attention deficit hyperactivity disorder from late childhood to early adolescence. *Journal of Child Psychology and Psychiatry, 46,* 824-836.

Egeland, J. (2007). Differentiating attention deficit in adult ADHD and first episode schizophrenia. *Archives of Clinical Neuropsychology, 22,* 763-771.

Egeland, J. (2010). Frequency of attention deficit in first episode schizophrenia compared to ADHD. *Applied Neuropsychology, 17,* 125-134.

Egeland, J.; Johansen, S.N. & Ueland, T. (2011). Central processing energetic factors mediate impaired motor control in ADHD combined subtype but not in ADHD inattentive subtype *Journal of Learning Disabilities,* Published online before print June 17, 2011, doi: 10.1177/0022219411407922

Egeland, J.; Johansen, S.N. & Ueland, T. (2010). Do low effort learning strategies mediate impaired memory in ADHD? *Journal of Learning Disabilities, 43,*430-440

Egeland, J.; Johansen, S.N, Ueland, T. (2009). Differentiating between ADHD sub-types on CCPT measures of sustained attention and vigilance. *Scandinavian Journal of Psychology,50,* 347-354.

Egeland, J. & Kowalik-Gran, I. (2010a). Measuring several aspects of attention in one test: The factorstructure of Conners' Continuous Performance Test. *Journal of Attention Disorders,13,* 339-347.

Egeland, J. & Kowalik-Gran, I (2010b). Validity of the factor structure of Conners' CPT. *Journal of Attention Disorders, 13,*347-357.

Egeland, J., Rund, B.R., Sundet, K., Asbjørnsen, A., Landrø, N.I., Lund, A., Roness, A., Stordal, K.I., Hugdahl, K. (2003). Attention profile in schizophrenia and depression: Differential effects of processing speed, selective attention and fatigue. *Acta Psychiatrica Scandinavica, 108,* 276-284.

Egeland, J.; Sundet, K.; Rund, B.R.; Asbjørnsen, A.; Hugdahl, K.; Landrø, N.I.; Lund, A.; Roness A. & Stordal, K.I. (2003). Sensitivity and specificity for memory dysfunction in schizophrenia: A comparison with major depression. *Journal of Clinical and Experimental Neuropsychology, 25,* 79-93.

Ek, U; Westerlund, J.; Holmberg, K. & Fernell, E. (2011). Academic performance of adolescents with ADHD and other behavioural and learning problems: A population-based longitudinal study. *Acta Paediatrica, 100,*402-406.

Epstein, J.N.; Erkanli, A.; Conners, C.K.; Klaric, J.; Costello, J. E. & Angold, A. (2003). Relations between continuous performance test performance measures and ADHD behaviors. *Journal of Abnormal Child Psychology, 31,* 543-554.

Feredoes, E.; Tononi, G. & Postle, B.R. (2006). Direct evidence for a prefrontal contribution to the control of proactive interference in verbal working memory. *Proceedings of the National Academy of Science, 103,* 19530-19534.

Geurts, H. M.; Verte, S.; Oosterlaan, J.; Roeyers, H. & Sergeant, J. A. (2005). ADHD subtypes: do they differ in their executive functioning profile? *Archives of Clinical Neuropsychology, 20*(4), 457-477.

Havey, J.M., Olson, J.M., & McCormick, C. (2005). Teacher's perceptions of the incidence and management of Attention-Deficit Hyperactivity Disorder. *Applied Neuropsychology, 12,*120-127.

Helps, S.K.; Broyd, S.J.; James, C.J.; Karl, A.; Chen, W. & Sonuga-Barke, E.J.S. (2010). Altered spntaneous low frequency brain activity in Attention Deficit/Hyperactivity Disorder. *Brain Research, 1322,* 134-143.

Hervey, A.S.; Epstein, J.N. & Curry, J.F. (2004). Neuropsychology of adults with attention deficit/hyperactivity disorder: A meta –analytic review. *Neuropsychology, 18,* 485-503.

Johnstone, S.L.; Watt, A.L. & Dimoska, A. (2010). Varying required effort during interference control in children with AD/HD: Task performance and ERPs. *International Journal of Psychophysiology, 76,* 174-185.

Kim, H. (2011). Neural activity that predicts subsequent memory and forgetting: A meta-analysis of 74 fMRI studies. *Neuroimage, 54,*2446-2461.

Kuntsi, J.; Wood, A.C.; Frühling, R.; Johnson, K.A.; Andreou, P.; Albrecht, B. ... Asherson, P. (2010). Separation of cognitive impairments in Attention-Deficit/Hyperactivity Diosrder into 2 familial factors. *Archives of General Psychiatry, 67,*1159-1167.

Liddle, E.B.; Hollis, C.; Batty, M.J.; Groom, M.J.; Totman, J.J.; Liotti, M.; Sceri, G. & Liddle, P.F. (2011). Task-related default mode network modulation and inhibitory control in ADHD: effects of motivation and mythylphenidate. *Journal of Child Psychology and Psychiatry, 52,* 761-771.

Luman, M.; Oosterlaan, J.; Hyde, C.; van Meel, C. & Sergeant, J.A. (2007). Heart rate and reinforcement sensitivity in ADHD. *Journal of Child Psychology and Psychiatry, 48,* 890-898.

Luman, M.; Osterlaan, J. & Sergeant, J.A. (2008). Modulation of response timing in ADHD, effects of reinforcement valence and magnitude. *Journal of Abnormal Child Psychology, 36,* 445-456.

McDonald, C.; Bauer, R. M.; Grande, L.; Gilmore, R. & Roper, S. (2001). The role of the frontal lobes in memory: Evidence from unilateral frontal resections for relief of intractable epilepsy. *Archives of Clinical Neuropsychology, 16,* 571-585.

McGee, R.A.; Clark, S.E. & Symons, D.K. (2000). Does the Conners' CPT aid in ADHD Diagnosis? *Journal of Abnormal Child Psychology, 28,* 415-424.

Mirsky, A.F.; Anthony, B.J.; Duncan, C.C.; Ahearn, M.B. & Kellam, S.G. (1991). Analysis of the elements of attention: A neuropsychological approach. *Neuropsychology Review, 2,* 109-145.

Nigg, J.T.; Blaskey, L.G.; Huang-Pollock, C. & Rappley, M.D. (2002). Neuropsychological Executive functions and DSM-IV ADHD subtypes. *Journal of the American Academy of Child & Adolescent Psychiatry, 41,*59-66.

Pennington, B. F. & Ozonoff, S. (1996). Executive functions and developmental psychopathology. *Journal of Child Psychology and Psychiatry, and allied disciplines, 37,* 51-87.

Rescorla, L.A., Achenbach, T.M., Ivonova, M., Dumenci, L., et al., (2007). Behavior and Emotional Problems Reported by Parents of Children Ages 6-16 in 31 Societies. *Journal of Emotional and Behavioral Disorders, 15,* 130-142.

Root, R.W. II & Resnick, R.J. (2003). An update on the diagnosis and treatment of attention-deficit/hyperactivity disorder in children. *Professional psychology: Research and Practice, 34,* 34-41.

Salimpoor, V.N. & Desrocher,M. (2006). Increasing the utility of EF assessment of executive function in children. *Developmental Disabilities Bulletin, 34,* 15-42.

Solanto, M.V.; Etefia, K. & Marks, D.I. (2004). The utility of self report measures and the continuous performance test in the diagnosis of ADHD in adults. *CNS Spectrums, 9,* 649-659.

Sonuga-Barke, E.J.S. (2002). Psychological herogeneity of AD/HD – a dual pathway model of behaviour and cognition. *Behavioural Brain Research, 130,*29-36.

Sonuga-Barke, E.J.S.; Bitsakou, P. & Thompson, M. (2010). Beyond the Dual Pathway Model: Evidence for the dissociation om timing, inhibitory and delay-related impairments in Attention-Deficit/Hyperactivity Disorder. *Journal of the American Academy of Child & Adolescent Psychiatry, 49,*345-355.

Sonuga-Barke, E.J.S.; Taylor, E.; Sembi, S. & Smith, J. (1992). Hyperactivity and Delay Aversion-I. The effect of Delay on choice. *Journal of Child Psychology &Psychiatry, 33,* 387-398.

Sonuga-Barke, E.J.S.; Wiersma, J.R.; van der Meere, J.J. & Roeyers, H. (2010). Context-dependent dynamic processes in Attention Deficit/Hyperactivity Disorder: Differentiating common and unique effects of state regulation deficits and delay aversion. *Neuropsychology Review, 20,* 86-102.

Sergeant, J.A. (2005). Modeling Attention-Deficit/Hyperactivity Disorder: A critical appraisal of the cognitive-energetic model. *Biological Psychiatry, 57,* 1248-1255.

Sergeant,J.A.; Geurts, H.; Huijbregts, S.; Sheres, A. & Osterlaan, J. (2003). The top and the bottom of ADHD: a neuropsychological perspective. *Neuroscience and Biobehavioral Review, 27,* 583-592.

Sergeant, J.A.; Osterlaan, J. & van der Meere, J.J. (1999). Information processing and energetic factors in Attention Deficit/Hyperactivity Disorder. In Quay, H., & Hogan, A.E. (Eds) *Handbook of disruptive behavior disorders.* Kluwer Academic/Plenum Publishers, New York.

Stark, R.; Bauer, E.; Merz, C.J.; Zimmermann, M.; Reuter, M.; Plichta; M.M.; Kirsch, P.;Lesch, K.P.; Fallgatter, A.J.; Vaiti, D. & Herrmann, M.J. (2011). ADHD related behaviours are associated with brain activation in the reward system. *Neuropsychologica, 49,* 426-434.

Söderlund, G.B.W.; Sikström, S., Loftesnes, J.M. & Sonuga-Barke, E.J. (2010). The effects of background white noise on memory performance in inattentive school children. *Behavioral & Brain Functions, 6,* (www.behavioraland brainfunctions.com/content/6/55)

Toplak, M.E.; Bucciarelli, S.M.; Jain, U. & Tannock, R. (2009). Ececutive functions: Performance based measures and the Behavior Rating Inventory of Executive Function (BRIEF) in adolescents with Attention Deficit/Hyperactivity Disorder (ADHD). *Child Neuropsychology, 15,* 53-72.

Van der Meere, J. J. (2002). The role of attention. In Sandberg (Ed) *Hyperactivity and attention disorders of childhood.* Cambridge: Cambridge University Press.

Wasserstein, J. (2005). Diagnostic issues for adolescents and adults with ADHD. Journal of Clinical Psychology/In session, 61, 535-547.

Zelazo, P.D. & Müller, U. (2002). Executive function in typical and atypical development. In U.Goswani *Blackwell Handbook of Childhood Cognitive Development.* Blackwell Publishing: Oxford, UK.

Øie, M. & Rund, B.R. (1999). Neuropsychological Deficits in adolescent-onset schizophrenia compared with Attention Deficit Hyperactivity Disorder. *American Journal of Psychiatry, 156,* 1216-1222.

# Part 2

## ADHD in Applied Contexts

7

# Adolescent Academic Outcome of Childhood Attention-Deficit/Hyperactivity Disorder – A Population-Based Study

Kirsten Holmberg
*Department of Women and Children Health, Section for Paediatrics, Uppsala University, Sweden*

## 1. Introduction

Children with Attention-deficit/hyperactivity disorder (ADHD) (APA, 1994) or "subthreshold" ADHD (with symptoms not prominent enough to fulfil the diagnostic criteria; i.e., similar but milder ADHD problems) (AAP, 1997) may be at risk of developmental problems (Warner-Rogers et al., 2000) or impairment (e.g., peer rejection) (Hoza, 2007; Scahill et al., 1999). It has been suggested that the research should include examination of the whole range of severity of hyperactivity and inattention symptoms in the population. This may provide further information about the relationship between symptom severity and overall impairment, in order to further evaluate the relative risk of elevated ADHD symptoms (Warner-Rogers et al., 2000).

Children with ADHD or subthreshold ADHD have been found to achieve lower grades at school than their peers (Biederman et al., 1996; Loe & Feldman, 2007). ADHD has been reported to be associated with difficulties in overall cognitive functioning or in specific domains, in reading and mathematics, which are not merely directly associated with low IQ scores (Gillberg et al., 2004; Spencer & Biederman, 2007). Inattention and hyperactivity in young children may be correlated with poor long-term academic achievement (Daley & Birchwood, 2010). Early recognition of ADHD followed by effective interventions has a potential to improve the educational and social outcomes for the affected children (Rasmussen & Gilberg, 2000; Biederman & Faraone. 2005; Jones et al., 2008).

One aim in the present study was to assess if children in a Swedish community sample who show symptoms of inattention, hyperactivity, and impulsivity with or without formal diagnoses of ADHD at 10 years of age also show poor long-term school outcomes. Another aim was to explore what degree of inattentive and hyperactive symptoms during elementary school (age 7 and 10) cause school failure at age 16.

## 2. Participants and methods

### 2.1 Study population

This study was based on three data collections in the birth cohort of 1991 in Sigtuna, a municipality in Stockholm County with a total population of approximately 36 000 inhabitants (Figure 1). All schools in the municipality, including special education classes, participated in the data collection. The special education classes included children with intellectual disabilities or subnormal cognitive abilities ("slow learners"), autistic spectrum disorders or disruptive behaviour. The birth cohort comprised 536 children in 1998 when these children entered first grade at age seven. Data was collected from teachers and parents to 453 of these children (84.5%) in first grade. A second data collection was carried out among all children in fourth grade including children moving into the cohort between first and fourth grade (N=591) in 2001-2002. In wave 2 data was collected from teachers and parents of 92% (n=544) of which 422 children (79% of the entire population born 1991; 204 girls, 218 boys) had participated in grade one. The present study is based on the third wave of data collection completed at age 16 in grade 9 in 2007 when school results were obtained from the national register. Children for whom there was information from three data-sources—the parent and the teacher in grade one or grade four (Conners ratings) and the national register in ninth grade (final grades)—were included in the final study population.

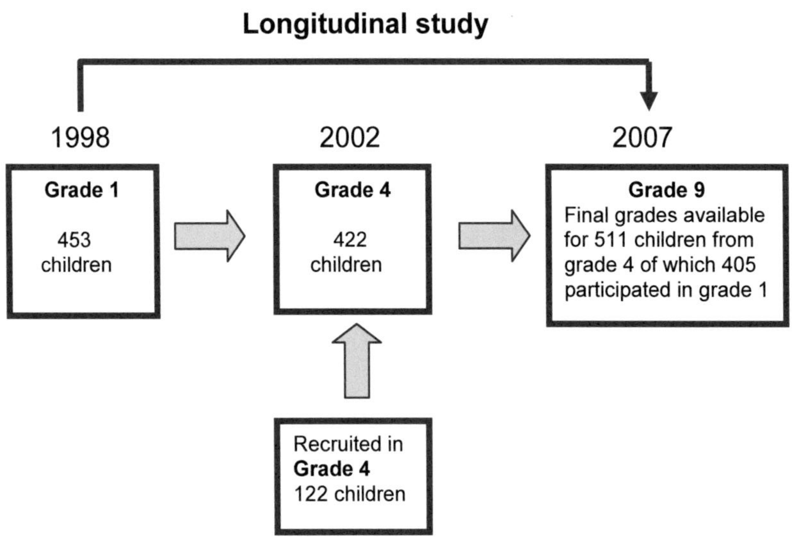


Fig. 1. Study design and participation.

Final grades at age 16 were available for 511 (87%) of the total population (N=591) in grade four. In 33 children participating in grade four no grades were available: 11 children had moved and had not received grades that were traceable in the national register, 4 had retaken one grade and were in eighth grade, 7 had joined a school for the mildly mentally retarded (IQ <71), 1 was already in upper secondary school, and 10 had failed to graduate. Of the total population of 536 first graders studied in 1998, it was possible to obtain grades for 405 (76%) in 2007.

Nonparticipants (i.e., school register data missing) in grade one (n=17) or in grade four (n=33) did not differ from participants by gender, socio-demographic conditions or mean Conners score reported by teacher. Neither was there any difference with respect to clinical ADHD-status in grade four. However, mean parental Conners score was higher in first ($p<0.001$) and fourth grade ($p<0.01$) among nonparticipants.

Ethical approval for the study in first and fourth grade was granted by the Ethics Committee at Karolinska Institutet, Stockholm and the follow-up study was approved by the Regional Ethics Committee in Uppsala.

## 2.2 Methods

### 2.2.1 Behavioural screening in first grade

At school entry into first grade the whole population was screened for developmental and behavioural problems by parental report in a questionnaire in connection with the routine health examination.

The questionnaire included the Conners 10-item scale (Conners, 1973; Conners, 1990a), and some questions about the socio-demographic characteristics of the household. The Conners 10-item scale is a commonly used well validated screening instrument for behavioural problems related to hyperactivity/impulsivity and emotional lability (Conners, 1973; Conners, 1990b). This scale consists of ten statements regarding the child's behaviour rated on a 4-point Likert scale, ranging from "0 - not at all true" to "3 - very much true" with a possible total score from 0 to 30 (Conners, 1990a). The scale is obtained from the 10 items constituting the Hyperactivity Index (HI) from the longer versions of the Conners scales (Goyette et al., 1978) and is also known as The Abbreviated Conners Rating Scales for parents (CPRS-HI) and teachers (CTRS-HI) (Conners, 1990a) and as the Abbreviated Symptom Questionnaire-Parent/Teacher (ASQ-P/T) (Conners, 1973; Conners, 1990b). A score of at least 10 has been recommended to identify attention deficits in a Swedish context (Landgren et al., 1996; Kadesjö & Gillberg, 1998), while a score of 15 or higher has been the standard for selecting children with hyperactivity at a level of clinical concern (Jones et al., 2008; Rowe & Rowe, 1997; Ullmann et al., 1985).

The parental version of the screening questionnaire included some questions about the socio-demographic characteristics of the household. The socio-demographic questions included sex of the child, maternal country of birth, and maternal educational level. Educational level was recorded in three categories: 9 years or less of basic education, more than 9 years of basic education but less than 3 years of university education, and 3 or more years of university education. Maternal country of birth was recorded as Sweden, other Nordic countries, other European countries, and the rest of the world.

Seven months into the school year, the same questionnaire was completed by the child's main teacher. Information about having an ADHD diagnosis from a physician at school entry was collected from the child's school health records. Results from the developmental screening have been reported in a previous article (Holmberg et al., 2010).

### 2.2.2 ADHD in fourth grade

The study population was followed up during the academic year 2001–2002 in grade four. The children were screened for ADHD in a two-step procedure which has been described more extensively in previous articles (Holmberg & Hjern, 2006; Holmberg & Hjern, 2008). In the first step, teachers and parents rated the children in a structured questionnaire in connection with a routine health examination. This questionnaire included the Conners 10-item scale, the same scale as in first grade. A cut-off score of at least 10 was used on this scale. The questionnaire also included the executive functions screening scale (EFSS) (Ek et al., 2004), a scale of problems related to concentration and problem solving developed for this study to improve the possibility to identify children with mainly attention problems. A cut-off score of 17 was used on this scale, which has a range of 0 to 51, as to obtain a total of

30% of screen positive children in the study population (Holmberg & Hjern, 2008). Having a score above the cut-off point on at least two of the four ratings made by the teachers and parents was considered screen-positive. Questions about the socio-demographic characteristics of the household were added to the parental version of the screening questionnaire for the children not participating in grade one.

Teachers also rated symptoms of ADHD on the ADHD rating scale—IV (DuPaul et al., 1998) based on the *Diagnostic and Statistical Manual of Mental Disorders, 4th edition* (DSM-IV) (APA, 1994) for ADHD as a part of the teacher questionnaire in grade four. This rating scale consists of 18 items (i.e., statements) which correspond to all 18 of the DSM-IV criteria: 9 items indicating inattention; 6 items pertaining to hyperactivity, and 3 items related to impulsivity. The child's behaviour is rated on each item on a 4-point Likert scale, ranging from "0 - not at all true" to "3 - very much true." Scores of 2 or 3 (indicating that a behaviour is present "often" or "very often") on individual items were considered to indicate the presence of ADHD symptoms, thereby creating dichotomised outcome variables for each statement. This scale has been validated (DuPaul et al., 1998; Merell & Tymms, 2001) and is widely used in Sweden for rating of ADHD severity (Diamantopoulou et al., 2005). The scale has demonstrated excellent interrater reliability, also when applied as a standardised interview schedule (Landgren et al., 1996; Thunström, 2002). The total score on each dimension of symptoms was calculated by adding up the scores for each of the 9 inattentive items and the 9 hyperactive/impulsive items.

Each teacher was interviewed by the author (KH) regarding learning and behaviour problems. Information from the teachers was received for all children in the study.

In a second step, 92% (130/141) of the screen-positive children in grade four underwent further clinical diagnostic assessments of ADHD based on the DSM-IV (APA, 1994), by an experienced child neurologist (KH). This evaluation included a clinical interview with structured information from the parents about ADHD symptoms in the home based on DSM-IV (APA. 1994) including the ADHD rating scale—IV (DuPaul et al., 1998), neurological examination of the child, and cognitive assessment according to the WISC III (Wechsler. 1999). The teacher score on the ADHD rating scale—IV was also included in the clinical evaluation.

Based on the clinical assessment, the children were classified into four categories; (1) "pervasive ADHD", children who met DSM-IV criteria for ADHD at home as well as at school; (2) "situational ADHD", children who fulfilled the criteria for ADHD in one setting only, either at home (home only ADHD) or at school (school only ADHD) (Mannuzza, 2002); (3) "subthreshold ADHD", children with four or five criteria for ADHD in one or two settings (AAP, 1997); (4) "no ADHD", all other children, including those who were not selected for clinical assessment.

Ten screen negative children were also clinically assessed and they were all included in the "no ADHD" group. Attention and hyperactivity symptoms in the 11 screen-positive children who did not participate in the clinical examination were assessed by information from parent and teacher questionnaires, teacher interviews, school nurses, and telephone interviews with parents. None of the 11 children who dropped out was judged to have severe behavioural or attention problems and were therefore included in the study population in the "no ADHD" group.

The prevalence of the complete (pervasive) ADHD syndrome in fourth grade was 5.7% (n=29, 2 girls and 27 boys), of which 25 had the combined type according to the DSM-IV criteria (1). Situational ADHD was present in another 6.9% (n=35, 9 girls and 26 boys).

According to the school health records, six boys had been assessed by a multidisciplinary team and received the diagnosis of ADHD at six to seven years of age before starting school. One percent (n=5) of the study population was treated with stimulants in grade four.

### 2.2.3 School achievement in fourth grade

The teacher questionnaire also included three items about the child's academic achievement in reading, writing and mathematics. The teacher rated the child's difficulties on each item on a 4-point Likert scale, ranging from "0 - not at all true" to "3 - very much true." Scores of 2 or 3 (indicating that a difficulty is present "often" or "very often") on individual items were considered to indicate the presence of learning problems thereby creating dichotomized outcome variables for each subject.

| Learning variables | Total (N=511) n (%) | No ADHD ( n= 415) n (%) | Subthreshold ADHD (n=32) n (%) | Situational ADHD (n=35) n (%) | Pervasive ADHD (n=29) n (%) |
|---|---|---|---|---|---|
| No learning difficulties | 417 (82) | 363 (87) | 22 (5) | 17 (4) | 15 (4) |
| Reading or writing difficulties | 58 (11) | 27 (47) | 6 (10)* | 14 (24)*** | 11 (19)*** |
| Mathematics difficulties | 65 (13) | 38 (58) | 6 (9) | 12 (19)*** | 9 (14)*** |

* $p < 0.05$, *** $p < 0.001$

Table 1. Learning difficulties according to teachers' ratings in children with attention-deficit/hyperactivity disorder (ADHD) in fourth grade.

No children were diagnosed with specific learning disabilities, i. e. dyslexia or dyscalculia at 10 years of age. Teachers reported 18% of fourth graders to have learning difficulties (Table 1). Eleven percent of the children had reading or writing problems while 13% had difficulties in arithmetical skills. These learning problems were strongly associated with situational and pervasive ADHD (p<0.001). In addition, reading or writing difficulties tended to be reported more frequent in children with subthreshold ADHD than those without ADHD (p<0.05) (Table 1).

### 2.2.4 School achievement in ninth grade

The Swedish school system is a 9-year compulsory school for children between 7 and 16 years of age. After finishing compulsory school, students receive an admission qualification, calculated from 16 subjects. The grade for each subject is defined as 0, 10, 15 or 20 scores. Therefore, the maximum total grade is 320 scores and is used as instrument of selection when applying to upper secondary school. In order to qualify for further studies in upper secondary school, a student needs to have attained certificate in core subjects: Swedish language, English and Mathematics. The grading level in each school is under national supervision by the Swedish School Authority through national tests in key subjects. The upper secondary school is divided into a national theoretical programme, giving authorization for university studies, and a national practical programme, leading more directly to work. In 2007, students with a minimum 151 grade point average or higher were admitted to the theoretical program at the local high school. Those with a score of at least 101 and passing grades in core subjects were accepted to the practical program.

In the present study, school grades from the National School Register, which is administered jointly by the Swedish National School Administration and Statistics Sweden, were used to calculate grade point average and qualification for further studies for all students. This register encompasses information on each individual's educational achievement that is grades by subject as well as grade point average. For this study, grades from five subjects were analysed: Swedish, English, Mathematics, History, Physics, Sports education and Music. The first five were analysed as examples of "theoretical subjects" and the two latter as "practical subjects". The register also encloses national tests results in three core subjects: Swedish, English and Mathematics for all students graduating from the ninth year. The national tests are carried out some weeks before the children graduate. The data from the National School Register are of high quality and summary statistics are published regularly (National School Register, 2010).

Information about special educational support in grade nine was collected by the author (KH) interviewing all teachers working in special education programs in grade nine.

#### 2.2.5 Statistical analysis

Chi- square, Fisher's exact test, and analysis of variance (ANOVA) were used to examine differences in relationship between ADHD symptoms in grade one and four, learning difficulties in grade four and school outcome variables in ninth grade. Associations of grade point average from the final ninth grade in the Swedish compulsory school system and qualification for upper secondary school with pervasive ADHD status in grade four were tested in linear and logistic regression models, respectively. Individuals with an incomplete course were excluded from the analysis of that particular course. Model 1 was crude while Model 2 was adjusted for sex and maternal education. In Model 3, 4 and 5, we added variables considered as learning difficulties. Results were expressed as B-coefficients with 95% confidence intervals (95% CI) in the linear regression analysis and OR with 95% CI in the logistic regression analysis. All statistical analyses were carried out using the SPSS 17.0 software package for Windows.

## 3. Results

### 3.1 Bivariate group outcome comparisons

The socio-demographic characteristics of the children in the study and learning difficulties in grade four by academic outcomes in grade nine are presented in Table 2. The overall mean grade was lower for boys than girls ($p<0.001$). There was no gender difference for being qualified for further studies in upper secondary school. Impaired academic achievement in ninth grade was more common among children from households where the parents had short education compared to households where parents had a university education (p<0.01) (Table 2). Grade point average was lower and not being qualified for upper secondary school was more frequent among children with previous learning difficulties compared to those with no reported difficulties in grade four (p<0.001) (Table 2).

Children with a at least one Conners score in parental report at age 10 or in teacher's report at age 7 or 10 had lower mean grade and increased prevalence of not qualifying for further studies at age 16 (p<0.001) (Table 3) than children with no reported hyperactivity. The cut-off score of 5 in parental report in first grade was related to impaired educational outcome in grade nine (p<0.01). At least one inattentive or one hyperactive symptom according to teachers' ratings on the ADHD symptom scale in grade four was associated with lower grade at 16 years of age (p<0.001) (Table 3).

| Measures | N | **Grade point average[1]** | | | **Qualified for upper secondary school** | **Not qualified[2] for upper secondary school** |
|---|---|---|---|---|---|---|
| | | mean | SD | 95% C.I. | n (%) | n (%) |
| **Socio-demographic variables** | | | | | | |
| Sex | | | | | | |
| Boys | 268 | 194.89*** | 60.32 | 187.63 - 202.14 | 234 (87) | 34 (13) |
| Girls | 243 | 213.74 | 65.49 | 205.47 - 222.02 | 218 (90) | 25 (10) |
| Maternal education | | | | | | |
| 0-9 years | 128 | 174.10*** | 61.59 | 163.33 - 184.87 | 107 (84) | 21 (16)** |
| 10-12 years | 286 | 203.27*** | 60.80 | 196.19 - 210.35 | 252 (88) | 34 (12) |
| 13 + years | 97 | 244.85 | 50.34 | 234.70 - 254.99 | 93 (96) | 4 (4) |
| Country of birth of mother | | | | | | |
| Sweden | 397 | 204.06 | 64.92 | 197.65 - 210.46 | 353 (89) | 44 (11) |
| Other Nordic countries | 26 | 209.42 | 71.82 | 180.41 - 238.43 | 22 (85) | 4 (15) |
| Other European countries | 13 | 228.46 | 59.81 | 192.32 - 264.60 | 12 (92) | 1 (8) |
| Rest of world | 75 | 196.60 | 52.20 | 184.59 - 208.61 | 65 (87) | 10 (13) |
| **Learning variables** | | | | | | |
| No learning difficulties | 417 | 213.68 | 61.75 | 207.74 - 219.62 | 389 (93) | 28 (7) |
| Reading or writing difficulties | 58 | 165.17*** | 48.77 | 152.35 - 178.00 | 41 (71) | 17 (29)*** |
| Mathematics difficulties | 65 | 155.08*** | 51.41 | 142.34 - 167.81 | 40 (62) | 25 (38)*** |

[1]Overall grade point average: possible range 0-320.
[2]Not qualified for upper secondary school; i. e. not receiving passing grades in Swedish, English and Mathematics.
** p < 0.01, *** p < 0.001

Table 2. Socio-demographic variables and learning difficulties in fourth grade in relation to school performance in grade nine.

| | | Grade point average | | | Qualified for upper secondary school | | Not qualified for upper secondary school | |
|---|---|---|---|---|---|---|---|---|
| **Measures** | n | mean | SD | 95% C.I. | n | (%) | n | (%) |
| **Grade one** | (n=405) | | | | (n=354) | | (n=51) | |
| *Conners* | | | | | | | | |
| Parent 0 | 145 | 209.66 | 65.64 | 198.88 – 220.43 | 129 | (89) | 16 | (11) |
| Parent 1-30 | 260 | 200.12 | 62.54 | 192.48 – 207.75 | 226 | (87) | 34 | (13) |
| Parent 5-30 | 99 | 186.97** | 68.01 | 173.40 – 200.53 | 82 | (83) | 17 | (17) |
| Parent 10-30 | 31 | 172.58** | 71.09 | 146.51 – 198.65 | 23 | (74) | 8 | (26)* |
| Parent 15-30 | 11 | 152.27** | 59.60 | 112.23 – 192.31 | 7 | (64) | 4 | (36)* |
| Teacher 0 | 187 | 217.89 | 58.97 | 209.38 – 226.40 | 173 | (92) | 14 | (8) |
| Teacher 1-30 | 218 | 191.22*** | 65.23 | 182.51 – 199.92 | 182 | (84) | 36 | (16)** |
| Teacher 5-30 | 99 | 173.69*** | 68.06 | 160.11 – 187.26 | 73 | (74) | 26 | (26)*** |
| Teacher 10-30 | 49 | 155.82*** | 64.46 | 137.30 – 174.33 | 32 | (65) | 17 | (35)*** |
| Teacher 15-30 | 27 | 131.67*** | 61.75 | 107.24 – 156.10 | 14 | (52) | 13 | (48)*** |
| **Grade four** | (n=511) | | | | (n=450) | | (n=61) | |
| *Conners* | | | | | | | | |
| Parent 0 | 165 | 226.88 | 60.34 | 217.60 – 236.15 | 159 | (96) | 7 | (4) |
| Parent 1-30 | 346 | 192.88*** | 62.06 | 186.31 – 199.44 | 291 | (84) | 54 | (16)*** |
| Parent 5-30 | 159 | 178.30*** | 61.43 | 168.68 – 187.92 | 128 | (80) | 31 | (20)*** |
| Parent 10-30 | 59 | 155.51*** | 60.34 | 139.79 – 171.23 | 44 | (75) | 15 | (25)*** |
| Parent 15-30 | 33 | 150.76*** | 63.84 | 128.12 – 173.39 | 21 | (64) | 12 | (36)*** |
| Teacher 0 | 244 | 228.85 | 60.15 | 221.27 – 236.44 | 231 | (95) | 13 | (5) |
| Teacher 1-30 | 267 | 181.01*** | 57.65 | 174.07 – 187.96 | 219 | (82) | 48 | (18)*** |
| Teacher 5-30 | 130 | 178.77*** | 57.61 | 168.77 – 188.77 | 103 | (79) | 27 | (21)*** |
| Teacher 10-30 | 75 | 166.60*** | 57.25 | 153.43 – 179.77 | 55 | (73) | 20 | (27)*** |
| Teacher 15-30 | 48 | 161.04*** | 51.48 | 146.09 – 175.99 | 36 | (75) | 12 | (25)** |
| ***Teacher ratings of DSM-IV criteria in grade four*** | (n=511) | | | | (n=450) | | (n=61) | |
| Inattentive score 0 | 363 | 218.56 | 57.48 | 212.64 – 224.48 | 338 | (93) | 25 | (7) |
| Inattentive score 1-9 | 148 | 167.09*** | 63.02 | 156.78 – 177.40 | 112 | (76) | 36 | (24)*** |
| Inattentive score 6-9 | 47 | 152.17*** | 49.02 | 137.62 – 166.73 | 32 | (68) | 15 | (32)*** |
| Hyperactivity score 0 | 416 | 211.06 | 62.41 | 205.04 – 217.07 | 374 | (90) | 42 | (10) |
| Hyperactivity score 1-9 | 95 | 172.32*** | 58.57 | 160.38 – 184.25 | 76 | (80) | 19 | (20)** |
| Hyperactivity score 6-9 | 31 | 163.55*** | 41.07 | 148.48 – 178.61 | 24 | (77) | 7 | (23) |

* p < 0.5, ** p < 0.01, *** p < 0.001

Table 3. Academic outcomes at end of grade nine by childhood ADHD-symptoms in first and fourth grade.

All three ADHD-groups were associated with impaired school outcome (Table 4). Thirty-five percent of children with pervasive ADHD in grade four did not qualify for upper secondary school compared with 8% of those without ADHD ($p<0.001$). The corresponding prevalence for children with subthreshold or situational ADHD was 25% and 26%, respectively. Adolescents with pervasive ADHD in childhood differed from unaffected peers in all educational outcomes except grade in History (Table 4). Situational ADHD showed a similar distribution of low grades at 16 years of age except in the subjects in Swedish, English and Sports education. The subthreshold and the pervasive ADHD group required special educational support in ninth grade more often than the no ADHD or situational ADHD group ($p<0.05$).

| **Academic variables** | Total (N=511) n (%) | No ADHD ( n= 415) n (%) | Subthreshold ADHD (n=32) n (%) | Situational ADHD (n=35) n (%) | Pervasive ADHD (n=29) n (%) |
|---|---|---|---|---|---|
| Grade point average (mean) | 203.86 | 214.45 | 164.69*** | 157.86*** | 151.03*** |
| Not qualified for upper secondary school | 61 (12) | 34 (8) | 8 (25)** | 9 (26)*** | 10 (35)*** |
| Not qualified for practical programme at upper sec school | 28 (6) | 14 (3) | 3 (9) | 7 (20)*** | 4 (14)** |
| Not qualified for theoretical programme at upper sec school | 86 (17) | 49 (12) | 11 (34)*** | 14 (40)*** | 12 (41)*** |
| Repeated grade (are in grade 8) | 24 (5) | 11 (3) | 4 (13)* | 4 (11)* | 5 (17)*** |
| Special educational support during grade 9 | 63 (12) | 44 (11) | 7 (22)* | 5 (14) | 7 (24)* |
| No certificate in | | | | | |
| Swedish | 15 (3) | 6 (1) | 3 (9)* | 1 (3) | 5 (17)*** |
| English | 31 (6) | 16 (4) | 5 (16)** | 4 (11) | 6 (21)*** |
| Mathematics | 47 (9) | 26 (6) | 7 (22)*** | 6 (17)* | 8 (28)*** |
| History | 30 (6) | 18 (4) | 4 (13) | 5 (14)* | 3 (10) |
| Physics | 56 (11) | 33 (8) | 10 (31)*** | 7 (20)** | 6 (21)** |
| Sports | 32 (6) | 18 (4) | 6 (19)*** | 4 (11) | 4 (14)* |
| Music | 43 (8) | 23 (6) | 3 (9) | 7 (20)*** | 10 (35)*** |

* $p < 0.5$, ** $p < 0.01$, *** $p < 0.001$ ADHD, attention-deficit/hyperactivity disorder

Table 4. Academic outcomes at end of grade nine in children with attention-deficit/hyperactivity disorder (ADHD) in fourth grade.

Figure 2 summarizes the national test results. Children with pervasive ADHD in grade four tended to fail in all three subjects ($p<0.01$ – $p< 0.001$) while those with situational ADHD were unsuccessful in Swedish and mathematics more often than other children ($p<0.01$) (Figure 2). Not passing the English test was more common in subthreshold and pervasive ADHD ($p<0.01$). Subthreshold ADHD was also associated with impaired results in mathematics ($p<0.05$).

### 3.2 Multivariate outcome prediction

Table 5 presents the linear regression models of mean grade point average and pervasive ADHD. The first model is crude, the second is adjusted for sex and maternal education; model 3 - 5 are also adjusted for reading or writing difficulties, mathematics problems or

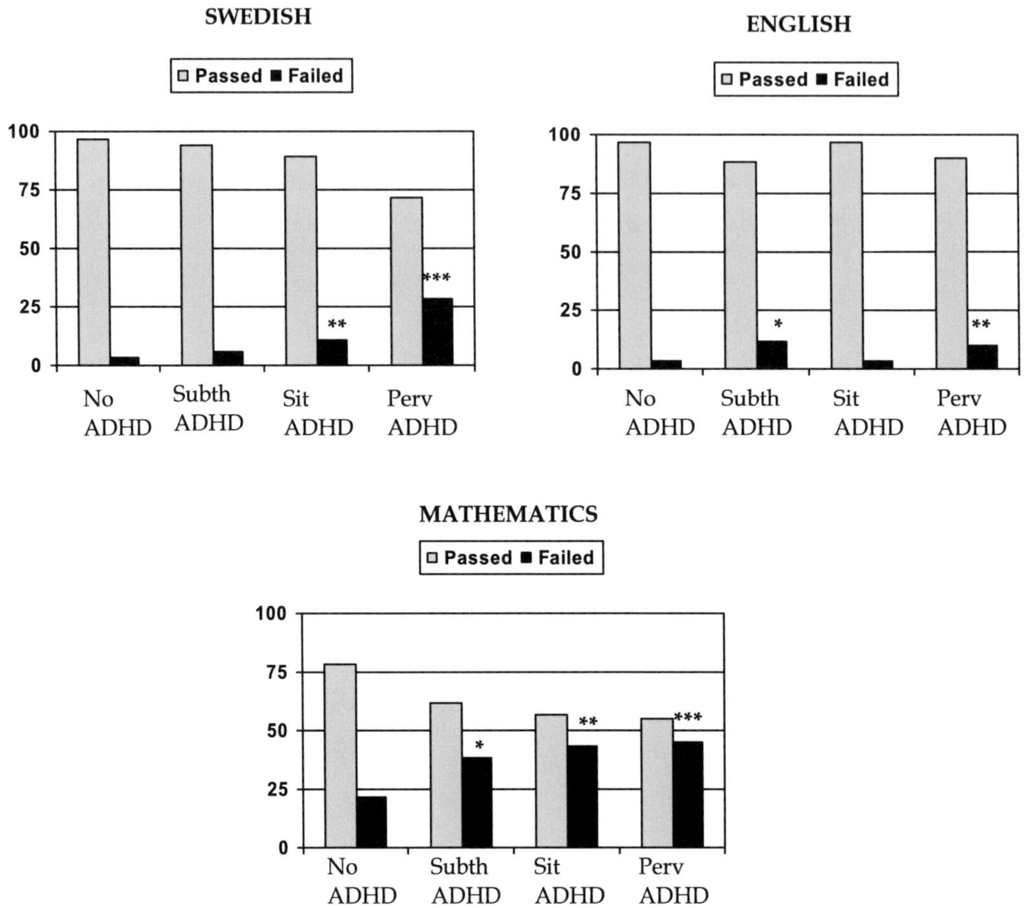


* p < 0.5, ** p < 0.01, *** p < 0.001

Fig. 2. National test results in Swedish, English and Mathematics in grade nine in children with attention-deficit/hyperactivity disorder (ADHD) in fourth grade.

both. In model 1, children with pervasive ADHD in grade four had a significantly lower overall mean grade than peers with no ADHD. In all adjusted models, the change in mean grade point average associated with ADHD was less marked, but still significant.

In Table 6 the risk of not being qualified for upper secondary school is presented in 5 models, adjusted for the same variables as in the linear regression analysis (Table 5). Pervasive ADHD at 10 years of age increased the odds of not qualifying for further studies at age 16 in all models (unadjusted odds ratio [OR] of 4.47; 95% confidence interval [CI]: 1.97 - 10.13). This estimate decreased slightly in the adjusted models. Of the socio-demographic covariates, only low maternal education had a significant relationship in model 2, raising the odds of poor school outcome for the child (OR: 1.78; 95% CI: 1.00 - 3.17). When learning difficulties were accounted for in model 3 - 5, the risk figure for pervasive ADHD was attenuated, but still significant. Children with problems learning mathematics in fourth grade had the highest risk of academic failure at end of compulsory school (OR: 6.46; 95% CI: 3.48 - 11.99).

| | Model 1 GPA (95% CI) | Model 2 GPA (95% CI) | Model 3 GPA (95% CI) | Model 4 GPA (95% CI) | Model 5 GPA (95% CI) |
|---|---|---|---|---|---|
| Pervasive ADHD | -56.00 (-79.36 – -32.64) | -47.99 (-70.87 – -25.11) | -39.92 (-62.87 – -16.98) | -38.19 (-60.40 – -15.97) | -35.71 (-58.11 – -13.31) |
| No ADHD | 0 | 0 | 0 | 0 | 0 |
| Sex: male | - | -12.96 (-23.56 – -2.35) | -10.53 (-21.06 – .01) | -13.58 (-23.78– -3.37) | -12.47 (-22.75 – -2.18) |
| Sex: female | - | 0 | 0 | 0 | 0 |
| Maternal education: low | - | -38.17 (-50.16 – -26.19) | -37.23 (-49.10 – -25.43) | -36.39 (-47.93– -24.85) | -36.18 (-47.71 – 24.65) |
| Maternal education: high | - | 0 | 0 | 0 | 0 |
| Reading or writing difficulties | - | - | -32.56 (-49.17 – -15.95) | - | -14.02 (-31.57 – 3.53) |
| No reading or writing difficulties | - | - | 0 | - | 0 |
| Mathematics difficulties | - | - | - | -47.71 (-64.84 – -34.58) | -44.63 (-61.02 – -28.24) |
| No mathematics difficulties | - | - | - | 0 | 0 |
| $R^2$ for model | .042 | .122 | .147 | .188 | .192 |

Model 1 is crude. ADHD, attention-deficit/hyperactivity disorder
Model 2 is adjusted for sex and maternal education.
Model 3, as Model 2 additionally adjusted for difficulties in reading or writing in grade 4.
Model 4, as Model 2 additionally adjusted for difficulties in mathematics in grade 4.
Model 5, as Model 2 additionally adjusted for difficulties in reading, writing and mathematics in grade 4.
CI, confidence interval.

Table 5. Linear regression of pervasive ADHD and mean grade point average (GPA) (N=511).

# 4. Discussion

## 4.1 Discussion of results

### 4.1.1 ADHD symptoms

This population-based study demonstrates that symptoms of inattention at age 7 and 10 as well as clinically diagnosed subtreshold ADHD and ADHD in 10-year-olds are associated with lower grade point average at the age of 16 and not being qualified for upper secondary school. Both levels of childhood ADHD are correlated to grade retention and not passing national tests in core subjects in grade nine. There appears to be a gradient relationship

| | Model 1<br>OR (95% CI) | Model 2<br>OR (95% CI) | Model 3<br>OR (95% CI) | Model 4<br>OR (95% CI) | Model 5<br>OR (95% CI) |
|---|---|---|---|---|---|
| Pervasive ADHD | 4.47<br>(1.97 - 10.13) | 4.40<br>(1.86 - 10.42) | 3.38<br>(1.38 - 8.30) | 3.29<br>(1.30 - 8.28) | 3.11<br>(1.22 - 7.93) |
| No ADHD | 1 | 1 | 1 | 1 | 1 |
| Sex: male | - | .99<br>(.56 - 1.75) | .89<br>(.49 - 1.59) | 1.04<br>(.57 - 1.89) | .99<br>(.54 - 1.82) |
| Sex: female | - | 1 | 1 | 1 | 1 |
| Maternal education: low | - | 1.78<br>(1.00 - 3.17) | 1.72<br>(.96 - 3.08) | 1.73<br>(.95 - 3.17) | 1.72<br>(.95 - 3.16) |
| Maternal education: high | - | 1 | 1 | 1 | 1 |
| Reading or writing difficulties | - | - | 3.10<br>(1.56 - 6.16) | - | 1.44<br>(.65 - 3.18) |
| No reading or writing difficulties | - | - | 1 | - | 1 |
| Mathematics difficulties | - | - | - | 6.46<br>(3.48 - 11.99) | 5.66<br>(2.86 - 11.20) |
| No mathematics difficulties | - | - | - | 1 | 1 |
| $R^2$ for model | .021 | .028 | .046 | .087 | .088 |

Model 1 is crude. ADHD, attention-deficit/hyperactivity disorder
Model 2 is adjusted for sex and maternal education.
Model 3 is adjusted for sex, maternal education and reading or writing difficulties in grade 4.
Model 4 is adjusted for sex, maternal education and mathematics difficulties in grade 4.
Model 5 is adjusted for sex, maternal education and difficulties in reading, writing and mathematics in grade 4.
CI, confidence interval.

Table 6. Logistic regression of pervasive ADHD in the fourth grade and not being qualified for upper secondary school (N=511).

between the number of symptoms and the frequency of subsequent adverse outcome. The risk of not being qualified for further studies after ninth grade is about three times higher in children diagnosed with pervasive ADHD than in other children. Children with learning difficulties according to teacher's reports at age 10, also have increased risk of academic underachievement. Low mathematical ability is a strong predictor of educational underachievement in 16-year-olds and moderates the effect of ADHD on school outcome although the association remains significant.

Our results replicate findings of Bussing et al. that childhood ADHD and subthreshold ADHD herald significant risks for lower educational achievement in adolescence (Bussing et al., 2010). In contrast to the previous study carried out in the US, both subthreshold and pervasive ADHD were associated with increased grade retention despite both groups receiving special educational support in ninth grade more often than children with no

ADHD (Table 4). This may reflect the possibility for children absent of a full ADHD diagnosis to qualify for special school services in the Swedish educational system. However, despite receiving extra academic support, , overall mean grade tended to be lower in in pervasive ADHD compared to children with subthreshold status and the prevalence of retaking one grade did not differ essentially between the subgroups (Table 4). Further studies with more detailed information about the remedial educational services both in special and general education classroom settings are warranted.

Inattentive/hyperactive symptoms in grade one and grade four defined as positive ratings on Conners Parent or Teacher Rating scales were associated with low grade point average and increased risk of not being qualified for upper secondary school. The cut-off level of at least one Conners score was related to negative school outcome (Table 3) in contrast to a threshold of 10 which is recommended to identify attention deficits in a Swedish context (Landgren et al., 1996; Kadesjö & Gillberg, 1998). Teacher's assessment on the ADHD rating scale—IV revealed that at least one inattentive or at least one hyperactive symptom according to the DSM-IV criteria at age 10 was correlated to not being qualified for further studies and a lower overall mean grade at age 16 (Table 3). Inattentiveness tended to result in slightly lower outcome results than hyperactivity. As for the Conners instrument the cut-off level on the DSM-IV symptom scale associated with underachievement at end of compulsory school was lower than the score of =>6 usually applied for identifying ADHD in epidemiological studies (Merell & Tymms, 2010). These results indicate that children with less behaviour problems than those at risk of developing clinically ADHD may also need specific educational attention. However, young children are frequently observed to be active and impulsive and this does not necessarily mean that they are not learning. According to previous research, the inattention element of ADHD appears to be the most important factor associated with underachievement in reading and mathematics (Diamantopoulou et al., 2007; Merell & Tymms, 2010). Further analyses in the present study population have revealed that Conners item 4 and 6 reflecting attention problems and Inattentive item 5 according to the DSM-IV criteria are the strongest predictors of poor school outcome (Holmberg & Bölte, 2011). These findings are in line with results from a longitudinal study by Breslau et al. demonstrating that symptoms of inattention at age 6 predict math and reading achievement at 17 years of age (Breslau et al., 2009).

Attention problems are likely to negatively influence children's academic achievement beginning in the early grades (Merrell & Tymms, 2001, Barbaresi et al., 2007). Students who have difficulties focusing on classroom activities or completing homework assignments because of their attention problems are likely to be less efficient learners compared with their classmates without attention problems. Inefficient learning in the early grades may limit students' ability to acquire basic skills that are necessary for developing higher level math and reading skills (Breslau et al., 2009). Learning problems in lower grades may cause additional inattentive behaviour and thereby further complicate the situation in school for students as they advance to the higher curricular demands of the later grades. In the longitudinal study by Barbaresi et al. following school-age children with ADHD into late adolescence, it was evident that the cumulative incidence of absenteeism and grade retention both increased as the children progressed from elementary school through high school (Barbaresi et al., 2007).

#### 4.1.2 Learning difficulties

Our findings confirm that ADHD and learning difficulties, especially in mathematics, are risk factors for poor school achievement. There is evidence showing that coexisting learning disabilities predicts further impaired academic outcomes for children with ADHD (Frazier et al., 2007; Bussing et al., 2010). When adjusting for ADHD and learning difficulties as predictors in regression analyses, it was evident that ADHD still had significant impact on academic outcome variables (Table 5 and 6). However, risk of coexisting attention and learning difficulties was not evaluated due to insufficient sample size.

#### 4.1.3 Medication

Whether stimulant medication in children and adolescents improves school performance or not has been discussed (Loe & Feldman. 2007; Barnard et al., 2010). Stimulant medication alone seems not to eliminate academic achievement deficits of ADHD, but may moderate the long-term academic outcome (Powers et al., 2008). The use of medication in our study population was low, only 1% of children diagnosed with ADHD were treated with stimulants at age 10. The prevalence of medical treatment may have increased to some extent during secondary school years, but this would probably not have had any major impact on the total result.

#### 4.1.4 Cognitive performance

Low school grades may also imply lower IQ. Children with ADHD show significant decreases in estimated full-scale IQ compared with controls but score on average within the normal range (Biederman et al., 1996; Gillberg et al., 2004, Daley & Birchwood, 2010). However, research that demonstrate the link between ADHD and academic underachievement have controlled for intelligence (Diamantopoulou et al., 2007) suggesting that individuals with ADHD perform academically at a lower level than would be predicted by their IQ. In a previous study in our cohort we have reported that children with ADHD performed better in the cognitive tests in connection with the clinical evaluation in fourth grade (according to age and gender-related norms for Swedish school children) than they did in terms of academic performance in grade nine (Ek et al., 2010). This result confirms the previous Swedish study by Diamantopoulou et al. indicating that children with ADHD underachieve academically in relation to their optimal cognitive capacity (Diamantopoulou et al., 2007).

#### 4.1.5 Early recognition of childhood ADHD symptoms

Our results stress the importance of early recognition of childhood ADHD and its subthreshold and situational presentations with or without coexisting reading, writing or mathematics difficulties. Young children with less prevalent symptoms of inattention may also be at risk of educational underachievement. Considering evidence that attention problems influence children's academic achievement negatively already in early grades, it may be of importance to start intervention during primary school to promote basic skills necessary for higher education. Early intervention may also prevent negative interaction with peers and teachers to evolve thereby reducing the increased risk of health complaints and increased risk of bullying behaviour in children with ADHD (Holmberg & Hjern, 2006; Holmberg & Hjern, 2008).

#### 4.1.5.1 Screening

Several different strategies may be considered in such interventions. In order to identify children at risk, screening at school entry has been recommended as part of school health surveillance for early detection of developmental or behavioural problems (Hall & Elliman, 2003; Swedish National Board of Health and Welfare, 2004). This may be the first step in secondary prevention in terms of social and educational support. Validated rating scales, such as a behaviour rating scale based on DSM-IV criteria may be applied (Merrell & Tymms, 2001; Holmberg, 2009). Teachers have been reported to underestimate ADHD symptoms and consider failure to persist in a task to be a sign of lack of interest, learning disability or family problems rather than inattention (Schachar & Tannock, 2006). Involving teachers in screening may be a way to increase their awareness of ADHD symptoms (Merell & Tymms, 2001; Holmberg, 2009), to discuss factors affecting the child's educational achievements and enhance the communication between teachers and the school health team. However, previous longitudinal studies of the developmental and behavioural screening of pre-schoolers and first graders in our cohort have demonstrated that the screening has low predictive values (15% – 50%) in relation to ADHD and school problems (Holmberg, 2009; Holmberg et al., 2010, Holmberg & Bölte, 2011). One reason for low efficiency of screening in young children may be that the behaviour of an individual child is influenced by many different factors that change over time. Changing family, teacher and peer relationships and increased demands on the child's intellectual capacity in the classroom in close interaction with the maturing brain create a dynamic context for the child's behaviour over time (Holmberg & Hjern, 2008). If screening for inattention and hyperactivity is carried out at the population level, it might be supplemented by a short clinical interview built into the routine school health programme. Such an approach may be more cost-effective and merit further evaluation (Holmberg, 2009).

#### 4.1.5.2 Early intervention

In the present study, about half of the children were reported to have at least one Conners score according to teachers in grade one or four and 29% of fourth graders had at least one inattentive score on the ADHD rating scale—IV in teacher's reports (Table 3). These results suggest that interventions should target all children, not only those with pronounced disruptive or inattentive behaviour. Several different strategies may be considered in such interventions. Interventions may target the family, the school situation or the child himself. Family-focused parent support programme with some evidence based support have been developed for children with ADHD symptoms in first grade (Jones et al., 2008; Sonuga-Barke et al., 2001). These programmes usually target all children with disruptive behaviour, not exclusively ADHD. Positive benefits from parent training, however, must be carefully balanced against the potential negative consequences of stigma associated with mislabeling. School focused interventions may include specific academic intervention strategies. Environmental modifications being offered to all students such as improvements in instruction/teaching methods, teaching materials, curriculum design, school physical designs, and leadership may also benefit children with ADHD-symptoms (Loe & Feldman, 2007). In addition, school may offer child focused interventions including behaviour management training, skill-based interventions (Breslau et al., 2009), emotional support and easy access to medication when indicated.

#### 4.1.5.3 Multi-disciplinary collaboration

Converging evidence regarding the importance of early childhood attention problems in predicting later school performance suggests that these problems should be a focus of concern across the multiple disciplines that address child health and well-being (Breslau et al., 2009). Early intervention services need to be supplemented with an effective strategy for identifying and supporting children who develop ADHD—or other neuropsychiatric disabilities that may interfere with learning—in the classroom when these problems arise. Such a strategy calls for close collaboration and communication between educators, who meet the children in the classroom every day, and the school health team (Holmberg et al., 2010). Children with school-related problems associated with ADHD require proper evaluation and treatment to prevent further impairment. Close collaboration between teachers and the school health team requires sufficient resources—both in terms of competence and finances—and may be an important ingredient in public health strategy for ADHD.

### 4.2 Limitations

Can the results of this study be generalized to all children with ADHD? Sigtuna is a medium-sized municipality with a population with a slightly more disadvantaged socioeconomic situation than the country as a whole, in terms of education, single parent household and the immigrant proportion of the population according to the Register of the Total Population and the Swedish Education Register. Thus, considering the higher rates of ADHD in families with low socioeconomic status (Swedish National Board of Health and Welfare, 2009) somewhat higher rates of ADHD compared with the national average should be expected in this study population. The school system is similar to the systems in most Swedish communities, with a preponderance of community-run schools with mainstream teaching methods. According to the National School Register (National School Register, 2010), 86.8% of children (n=4384) leaving compulsory school in grade nine in Sigtuna June 2007 were qualified for further studies and the mean grade point average was 203.3, compared to 89.1% (N=935 869) and 207.3, respectively, in the whole country. Thus, the associations between ADHD and impaired academic outcome reported in this study may be over-estimated compared to other societal contexts in Sweden.

Educational systems and school demands, mental health use and stimulant medication as well as the prevalence of ADHD varies considerably between countries. This suggests that the results of this study may, to a certain extent, be specific for Swedish schoolchildren.

Children being screen positive for ADHD-symptoms in grade four (and ten screen negative) were assessed by the clinician. The Conners 10-item scale has been validated in previous population-based studies in Sweden (Landgren et al., 1996; Kadesjö & Gillberg, 1998). The EFSS-scale has not been validated and we have no information about the sensitivity of this instrument which is a potential weakness in our study-design. Some children with ADHD may not have been correctly identified in fourth grade. Since the screening by questionnaires in grade four was completed by interviews of all teachers about ADHD-symptoms, it is unlikely that children with significant problems were not identified.

The method of data collection is the greatest strength of this study. Data about learning and/or behaviour problems in 2002 were collected within the school health system and used in connection with health visits to the school nurses and physicians. This explains the extraordinarily high participation rate of parents and teachers. Using the national register of final grades minimised the attrition rate since all schools in Sweden report to this register. We could retrieve grades for 41 children who had moved out of the study-population between first and second data collection. Another strength of the study is the use of multiple informants where data on behaviour is provided by both teachers and parents, data on socio-economic conditions by parents, and data on school results from national register.

Our sample is too small to allow for any conclusions to be drawn about gender differences in ADHD, whether girls with ADHD graduated from compulsory school with better academic results than boys with ADHD. Interaction effects of ADHD and learning difficulties on the outcome variables were not analysed due to the insufficient sample size.

## 5. Conclusions

This population-based study demonstrates a connection between mild as well as more severe ADHD-symptoms in young schoolchildren (age 7 and 10) and academic underachievement at 16 years of age. Schoolchildren with behavioural problems of inattention, hyperactivity and impulsivity but not reaching diagnostic threshold may nevertheless be at risk of impaired academic progress. The results suggest that subthreshold and situational ADHD deserve the same clinical attention and psychosocial treatment as pervasive ADHD to prevent further impairment. Children with learning problems, especially mathematics difficulties, in middle school with or without ADHD seem to be especially at risk of school failure. Close collaboration between health and educational personnel is required to identify and support children with attention and learning problems. A multi-disciplinary approach with integrated services may prevent further impairment. This needs to be further explored in large, prospective, longitudinal, and community-based studies. Future research on childhood ADHD symptoms and learning difficulties in larger populations with longer follow-up periods may reveal whether elevated ADHD symptoms only have the same impact on adverse outcome as coexisting problems. Finally, additional research is required to determine which pharmacologic, behavioural, and educational interventions can improve academic outcomes of children with ADHD.

## 6. Acknowledgements

We wish to thank Professor Anders Hjern of CHESS, the Centre for Health Equity Studies at Stockholm University, for supervising the design and execution of the study, ethical application and for valuable assistance to gain access to data from the national register. Financial support for this study has been provided by The Swedish Society of Medicine, the First of May Flower Annual Campaign, the Samariten Foundation and the Foundation Claes Groschinskys Memory. We thank the school authorities of Sigtuna, the school nurses, and the teachers, without whose assistance the study could not have been completed.

## 7. References

American Academy of Pediatrics (1997). *Diagnostic and Statistical Manual of Mental Disorders for Primary Care*. American Academy of Pediatrics, Washington DC, US.

American Psychiatric Association (1994). *Diagnostic and Statistical Manual of Mental Disorders,* 4 ed. American Psychiatric Association, Washington DC, US.

Barbaresi, W.J., Katusic, S.K., Robert, C., Colligan, R.C., Weaver, A.L. & Jacobsen, S.J. (2007). Long-Term School Outcomes for Children with Attention-Deficit/Hyperactivity Disorder: A Population-Based Perspective. *J Dev Behav Pediatr,* Vol.28(No.4), pp 265–73.

Barnard, L., Stevens, T., To, Y.M., Lan, W.Y. & Mulsow, M. (2010). The Importance of ADHD Subtype Classification for Educational Applications of *DSM-V. J Atten Disord,* Vol.13(No.6), 573–83. E pub 2009 Apr 16.

Biederman, J., Faraone, S., Milberger, S., et al. (1996). A prospective 4-year follow-up study of attention-deficit hyperactivity and related disorders. *Arch Gen Psychiatry,* Vol.53(No.5), pp 437–446.

Biederman, J. & Faraone, S.V. (2005). Attention-deficit hyperactivity disorder. *Lancet,* Vol.366(No.9481), pp. 237-248.

Breslau, J., Miller, E., Breslau, N., Bohnert, K.., Lucia, V. & Schweitzer J. (2009). The Impact of Early Behavior Disturbances on Academic Achievement in High School. *Pediatrics,* Vol.123(6), pp 1472-6.

Bussing, R., Mason, D.M., Bell, L., Porter, P. & Garvan, C. (2010). Adolescent outcome of childhood attention-deficit/hyperactivity disorder in a diverse community sample. *J Am Acad Child Adolesc Psychiatry,* Vol.49(No.6), pp 595-605. Epub 2010 May 1.

Conners, C.K. (1973). Rating scales for use in drug studies with children. *Psychopharmacology Bulletin* (Special issue, Pharmacotherapy of Children), pp. 24–29.

Conners, C.K. (1990a) *Manual for Conners Rating Scales.* Multi-Health Systems, Toronto, Canada.

Conners, C.K. (1990b). *Conners Abbreviated Symptom Questionnaire: Parent Version, Teacher Version Manual.* Multi-Health Systems, Toronto, Canada.

Daley, D. & Birchwood, J. (2010). ADHD and academic performance: why does ADHD impact on academic performance and what can be done to support ADHD children in the classroom? *Child Health Care Develop* Vol.36(No.4),pp 455-64. E pub 2010 Jan 13.

Diamantopoulou, S., Henricsson, L. & Rydell, A.M. (2005). ADHD symptoms and peer relations of children in a community sample: Examining associated problems, self-perceptions, and gender differences. *International Journal of Behavioral Development* Vol.29:388-398.

Diamantopoulou, S., Rydell, A., Thorell, L. B. & Bohlin, G. (2007). Impact of executive functioning and symptoms of attention deficit hyperactivity disorder on children's peer relations and school performance. *Developmental Neuropsychology,* Vol.32, pp 521–42.

DuPaul, G., Power, T., Anastopoulos, A. & Reid, R. (1998). *ADHD rating scale - IV. Checklists, norms, and clinical interpretation.* Guildford Press. New York, USA.

Ek, U., Holmberg, K., de Geer, L., Swärd, C. & Fernell, E. (2004). Behavioural and learning problems in schoolchildren related to cognitive test data. *Acta Paediatrica,* Vol.93(No.7), pp. 976- 981.

Ek, U., Westerlund, J., Holmberg, K. & Fernell, E. (2010). Academic performance in adolescents with ADHD and other behavioural and learning problems – a population-based longitudinal study. *Acta Paediatrica,*Vol.100(No.3),pp 402-6, Epub 2010 Nov 5.

Frazier, T.W., Youngstrom, E.A., Glutting, J.J. & Watkins, M.W. (2007). ADHD and achievement: meta-analysis of the child, adolescent, and adult literatures and a concomitant study with college students. *J Learn Disabil.* Vol.40: pp 49–65.

Gillberg, C., Gillberg, I.C., Rasmussen, P., Kadesjö, B., Söderström, H., Råstam, M., Johnson, M., Rothenberger, A. & Niklasson, L. (2004). Co-existing disorders in ADHD - implications for diagnosis and intervention. *Eur Child Adolesc Psychiatry,* Vol.13 Suppl 1:pp I80-92.

Goyette, C.H., Conners, C.K. & Ulrich, R.F. (1978). Normative data on revised Conners Parent and Teacher Rating Scales. *J Abnorm Child Psychol,* Vol.6(No.2), pp. 221-36.

Hall, D. & Elliman, D. (Eds) (2003) *Health for All Children,* (4th ed), Oxford University Press, Oxford, UK.

Holmberg, K. & Hjern, A. (2006). Health complaints in children with attention-deficit/hyperactivity Disorder. *Acta Paediatrica,* Vol.95(No.6), pp. 664-70.

Holmberg, K. & Hjern, A. (2008). Bullying and ADHD in 10 year-olds in a Swedish community. *Dev Med Child Neurol,* Vol.50(No.2), pp. 134-138, Epub 2007 Dec 19.

Holmberg K. (2009). *Health complaints, bullying and predictors of attention-deficit/hyperactivity disorder (ADHD) in 10-year-olds in a Swedish community* [dissertation], University of Uppsala, Uppsala, Sweden.

Holmberg, K., Sundelin, C. & Hjern, A. (2010). Routine developmental screening at 5.5 and 7 years of age is not an efficient predictor of attention-deficit/hyperactivity disorder at age 10. *Acta Paediatrica,*Vol. 99(No.1), pp. 112-120. Epub 2009 Sep 17.

Holmberg, K., & Bölte, S. (2011). Symptoms of ADHD at age 7 and 10 predict academic outcome at age 16. *Proceedings of the European Academy of Childhood Disability 23rd Annual meeting,* 8-11 June, 2011, Rome, Italy. Dev Med Child Neurol 2011 June;53(Suppl s3):60. Epub 2011 May 19.

Hoza, B. (2007). Peer functioning in children with ADHD. *J Pediatr Psychol* Vol.32, pp 655-63.

Jones, K., Daley, D., Hutchings, J., Bywater, T. & Eames, C. (2008). Efficacy of the Incredible Years Programme as an early intervention for children with conduct problems and ADHD: long-term follow-up. *Child: care, health and development,* Vol.34(No.3), pp. 380-90.

Kadesjö, B. & Gillberg, C. (1998). Attention deficit and clumsiness in Swedish 7-year-old children. *Dev Med Child Neurol,* Vol.40(No.12), pp. 796-804.

Landgren, M., Pettersson, R., Kjellman, B. and Gillberg, C. (1996). ADHD, DAMP and other neurodevelopmental/psychiatric disorders in 6-year-old children: epidemiology and co-morbidity. *Dev Med Child Neurol,* Vol.38(No.10), pp. 891-906.

Loe, IM. & Feldman, HM. (2007). Academic and educational outcomes of children with ADHD. *J Pediatr Psychol,* Vol.32(No.6), pp 643-54. Epub 2007 Jun 14. Review.

Mannuzza, S., Klein, R.G. & Moulton, J.L. 3rd (2002). Young adult outcome of children with "situational" hyperactivity: a prospective, controlled follow-up study. *J Abnorm Child Psychol* Vol.30:191-8.

Merrell, C. & Tymms, P. B. (2001). Innatention, hyperactivity and impulsiveness: their impact on academic achievement and progress. *Br J Edu Psychol,* Vol.71, pp 43–56.

National School Register (2010) (10 August 2011, date last accessed). URL: www.skolverket.se. The Swedish National Agency for Education, Stockholm, Sweden.

Powers, R.L., David J. Marks, D.J., Miller, C.J., Newcorn, J.H. & Halperin, J.M. (2008). Stimulant Treatment in Children with Attention-Deficit/Hyperactivity Disorder Moderates Adolescent Academic Outcome. *J Child Adolesc Psychopharmac,* Vol.18(No.5), pp 449-59.

Rasmussen, P. & Gillberg, C. (2000). Natural outcome of ADHD with developmental coordination disorder at age 22 years: a controlled, longitudinal, community- based study. *J Am Acad Child Adolesc Psychiatry,* Vol.39(No.11):1424-31.

Rowe, K.S. & Rowe, K.J. (1997).Norms for Parental Ratings on Conners' Abbreviated Parent-Teacher Questionnaire: Implications for the Design of Behavioral Rating Inventories and Analyses of Data Derived from Them. *J Abnormal Child Psychol,* Vol.25, pp. 425-451.

Scahill, L., Schwab-Stone, M., Merikangas, K.R., Leckman, J.F., Zhang, H. & Kasl, S. (1999). Psychosocial and clinical correlates of ADHD in a community sample of school-age children. *J Am Acad Child Adolesc Psychiatry* Vol.38:976-84.

Schachar, R. & Tannock, R. (2006). Syndromes of Hyperactivity and Attention Deficit. In: *Child and Adolescent Psychiatry,* (4th ed) (eds M. Rutter, E. and E. Taylor), pp. 399-418. Blackwell Science Ltd; Oxford, UK.

Sonuga-Barke, E.J., Daley, D., Thompson, M., Laver-Bradbury, C. & Weeks, A. (2001). Parent-based therapies for preschool attention-deficit/hyperactivity disorder: a randomized, controlled trial with a community sample. *J Am Acad Child Adolesc Psychiatry,* Vol.40(No.4), pp. 402-408.

Spencer, T.J., Biederman, J. & Mick, E. (2007): Attention-deficit/hyperactivity disorder: diagnosis, lifespan, comorbidities, and neurobiology. *J Pediatr Psychol* Vol.32(No.6), pp 631-42.

Swedish National Board of Health and Welfare (2004). *Socialstyrelsens riktlinjer för skolhälsovården. Rekommendationer för planering/tillsyn/metodutveckling.* [Manual for School Health Services]. Socialstyrelsen, Stockholm, Sweden, (in Swedish).

Swedish National Board of Health and Welfare (2009). Folkhälsorapporten. [Health in Sweden - The National Public Health Report 2005]. Socialstyrelsen, Stockholm, Sweden (in Swedish).

Thunström, M. (2002): Severe sleep problems in infancy associated with subsequent development of attention-deficit/hyperactivity disorder at 5.5 years of age. *Acta Paediatr* Vol.91, pp 584-92.

Ullmann, R.K., Sleator, E.K. & Sprague, R.L. (1985). A change of mind: the Conners abbreviated rating scales reconsidered. *J Abnorm Child Psychol,* Vol.13(No.4), pp. 553-565.

Warner-Rogers, J., Taylor, A., Taylor, E. & Sandberg, S. (2000). Inattentive behavior in childhood: epidemiology and implications for development. *J Learn Disabil* Vol.33, pp 520-36.

Wechsler D. (1999). *Wechsler intelligence scale for children* (3rd ed), Revised Psychological Corporation, New York, US.

# 8

# Remedial Education for Children with ADHD in Sweden

Jane Brodin
*Stockholm University*
*Sweden*

## 1. Introduction

The purpose of this chapter is to increase the understanding and knowledge of children with Attention Deficit Hyperactivity Disorder (ADHD) and their everyday lives in educational settings. As these children constitute an ever increasing group all over the world and the disorder really creates huge problems for them in school, this chapter will focus on remedial education in Sweden. In this chapter I will highlight the interdisciplinary research project Basic skills, social interaction and training of the working memory (BASTA) that included vital aspects related to education. Three research teams: one medical and two pedagogical were involved. The pedagogical teams at Stockholm Institute of Education (teacher training university) were responsible for child studies and special education. The tasks for the special education team was to study the basic academic skills and for the child studies the focus was on issues related to children's perspectives of the meaning of being a child with ADHD and on children's school environments. Issues related to education of children with ADHD are based on self-concept, teaching, ethical issues and children's views, as these aspects are important in school. I will also make some recommendations for education.

ADHD is a neuro-psychiatric disorder of childhood that is characterized by developmentally inappropriate levels of hyperactivity, impulsivity and inattentiveness. The prevalence of ADHD in Sweden is estimated to 3 to 6 per cent of all school-age pupils (Ljungberg, 2001; Ljusberg, 2009), and in the US the figures are similar, i.e. 3 to 7 per cent (DuPaul & White, 2006). Ljungberg (2001) argues that if also Deficit in Attention, Motor Control and Perception (DAMP) is included, the figure would increase to approximately 10 per cent. The disorder occurs more often in males than females with the sex ratio being about 3.4 to 1. Nevertheless, ADHD is one of the most common psychiatric disorders in childhood (Wells, 2004). The reason for the disorder is according to medical studies a deviation in the brain which leads to behavioural disturbances and social difficulties (Westerberg, 2004). Based on these findings the medical discourse has to a great extent influenced the Swedish educational system as well as the school systems in other countries (Harwood, 2006; Lloyd, 2006).

The number of detected children with ADHD is thus increasing and for these children and people in their immediate environment neuro-psychiatric disorders cause huge problems with regard to learning, teaching and social interaction. This has resulted in an increasing demand for special classes in school and we can today note that there are special classes for children with for instance ADHD, Asperger syndrome, dyslexia/reading and writing deficits, speech and language disorders, slow learners and for children with psychosocial problems. This trend is global in spite of the fact that most countries for the time being are promoting inclusion of all children in a school for all (Brodin, 2008; Norwich, 2008).

All children have the right to education, participation and equal opportunities according to the UN Convention on the Rights of the Child (1989) but for these children and their parents everyday life is often a struggle, as ADHD has not always been regarded as a real disorder. Many parents have felt unhappy and misunderstood and they have by people in the environment been regarded as incompetent to raise their children in a proper way (e.g., Hellström, 2004, 2010; Kadesjö, 2010). This has given many parents bad conscious and hard feelings of blame and guilt. This problem is well-known in research and in a guide for parents and professionals Chandler (2010) argues that despite the dramatic increase in the diagnosis and treatment of ADHD the disorder is not new. The topic has been highlighted in a number of reports, e.g. the International Consensus Statement on ADHD (Barkley, 2002). The consensus statement aims at confirming that ADHD is a real and valid disorder and that there is no doubt regarding its existence. Below is a brief summary of the report.

### 1.1 International consensus statement on ADHD (2002)

It appears from the report that ADHD is one of the most researched disorders in medicine. The statement is signed by 86 leading scientists; the majority professors from the medical field. The purpose was to stress the status of the diagnosis and to substantiate that ADHD is a medical condition but it is also a protest against inaccurate, populist stories of ADHD in media that may negatively influence children and adults who suffer from it to require professional help and treatment. The involved scientists realized that there was an obvious risk that thousands of sufferers would not 'seek treatment for their disorder', and added that 'there is no controversy regarding its existence' (p. 89). It appears that ADHD causes impairments in major life activities and that these children are at risk for physical injuries at a larger extent than other children, as they have difficulties to judge the consequences of an action. This is most reasonable a result of the psychological deficits in attention and inhibition. However, ADHD cannot be explained as a medical problem only as the environmental aspects (e.g. family or school setting) are contributing to the neuro-psychiatric condition of the child. The social interaction between the genetics and environmental factors are thus a decisive prerequisite, and both are of importance when the most suitable treatment is taken into consideration. The difficulties in everyday life primarily concern domains such as education, social relations, self-esteem, self-concept, family functioning, and independence (p. 90).

Furthermore it appears from the report that it is common that children with ADHD have other associated disorders mainly related to their social environment e.g., communication disorders and emotional disturbances. All these difficulties account for a large number of referrals to pediatricians, family physicians and child mental health professionals but less than half of all persons (primarily children) with ADHD receive treatment.

### 1.2 What do we know about ADHD?

ADHD is an ever increasing disorder among children especially in the western world and one reason for experiencing an increase is probably that we nowadays have opportunities to detect the disorder easier and have access to instruments to diagnose these children (Chandler, 2010). The disorder has been the focus of a large number of scientific and clinical studies over the last centuries and one of the front figures in Sweden is Gillberg (2005), who early noticed the difficulties the disorder caused children with DAMP and ADHD in school. Based on research he recommended placement in small groups, individual instructions in maths, reading and writing, short intervals of concentration during the first years in school, regular breaks and physical activities. His team also showed that these recommendations functioned in practice and made everyday life easier for the children. These recommendations are in fact supportive for all children, i.e. typical and atypical, but in school the situation often looks different. Too many pupils in each class are obstacles that make individual intervention problematic for the teachers. Consequently, many teachers feel insufficient and think that they are not good enough in their profession. Teachers often state in interviews that they would prefer to teach in another way than they do (Brodin, 2009) but the lack of time and too many pupils in each class is a reality in today's school.

Due to the inattentiveness and difficulties to maintain attention they need breaks and physical activities. The need to use their bodily capacities in combination with their often bully behaviour are challenging for the teachers and quite often the disorder in the classroom is extremely demanding. One way to meet the needs of these children is to be flexible and not necessarily keep to the time schedule and curricula. This might in many countries be controversial as teachers in general are supposed to follow the curricula. However, if a child's learning is the main goal in school this effort would contribute to better opportunities to take charge of the education.

What we really know from research is that ADHD is a disorder that creates many difficulties for the child and his or her family and the harm is noticed by the increased mortality, morbidity and difficulties in everyday life both for children and adults (Alin Åkerman, 2008; Ingvar, 2004; Teeter, 2004).

### 1.3 Tools for classification of ADHD

In order to classify the disorder different tools are used. The Diagnostic and Statistical Manual of Mental Disorders (DSM-IV) from 2000 and the Who's International Classification of Disorder (ICD-10) are the most frequently used tools to classify ADHD and it is evident that difficulties to maintain attention/concentration and hyperactivity/impulsiveness are the most common problems. Three different types of ADHD can be recognized: inattentiveness and hyperactivity mainly inattentiveness (AD), inattentiveness and hyperactivity mainly hyperactivity/impulsivity (HD), and inattentiveness and hyperactivity in combination (ADHD). A majority of the children have a combination of inattentiveness and hyperactivity.

From DSM-IV appears that the symptoms should have been noticed before the child is seven years old and appear within at least two areas (e.g., in school and at home) and to imply a clinical significant disability in daily life. The diagnostic criteria are defined, but still many professionals have difficulties to be sure as the degree of the problems differ in different environments and with different people around them. Ljusberg (2010) means that there are also certain situations and tasks in school that cause concentration difficulties. She thus supports Danby and Farrell (2004) who argue that when the school ascribes difficulties to pupils it contradicts its striving for children's agency, competence and participation. Ljusberg's standpoint is thus that the children acquire concentration difficulties in school (2010).

### 1.4 Research on children with ADHD

The majority of studies on children with ADHD focuses on medical, psychological, social and educational aspects. The school situation is often described as problematic. Barkley (1997) argues that the limited capacity to prevent the impulse from immediate action is the core in the ADHD complex of problems. He states that these children act immediately when they get an idea and they do not or cannot consider the consequences. For this reason they are often involved in conflicts with peers, parents and teachers. Many children with ADHD also have low self-esteem and self-confidence and fail in schoolwork (Brodin & Ljusberg, 2008; Ljusberg & Brodin, 2007).

The disorder can be defined and explained in many different ways depending on the researcher's perspective and knowledge. Most neurological studies highlight that persons with ADHD have low electrical activity in the brain and show less reactivity to stimulation in one or more regions in the brain (Klingberg *et al.*, 2005). The mounting evidence of neurological and genetic contributions to ADHD is a fact (Barkley, 2002). For this reason treatment with drugs that stimulates the central nervous system e.g., Dopamine, Retalin or similar drugs is often recommended, and this is one reason for critics e.g. from pedagogues and sociologists. Many doctors prescribe drugs to help the child and most researchers in neuro psychiatry in Sweden support medication (Eriksson & Ingvar, 2004; Kadesjö, 2010) but there has been an intensive debate for and against prescription of drugs for children with ADHD. However, with regard to all the negative consequences these children have to face in everyday life, drugs seem to be a solution for these children and their families. It is however vital to point out, that drugs is never the first and only choice when treatment is considered.

Nevertheless, ADHD leads to impairments in everyday life related to education, social relations and family functioning. The very first step is to look at the immediate environment in school and find out if it is possible to make any changes in the classroom that would facilitate learning and social interaction for these children. The physical environment in school can be adapted to better correspond to the needs of these children. Other factors of importance are teacher training, i.e. that the teachers have a formal teacher education, have high competence in children in need of special support, have knowledge of the difficulties children with ADHD have, and finally how they structure their teaching to make learning optimal. Equal value and consideration must be taken into account for teaching and learning.

From a pedagogical point of view Tannock and Martinussen (2001) stress that ADHD is primarily a cognitive disability, but like most researchers Ljungberg (2001) emphasizes that ADHD can develop differently and it is reasonable to believe that the disorder can be based both on genetic and psychosocial factors, or be a combination of both. Kadesjö (2010) stresses the connection between the prevalence of physical and emotional disturbances, and charging social factors. The higher prevalence of physical unhealth in children from socioeconomic disadvantaged families can be explained by an increased prevalence of risk factors. The environmental factors thus affect the disorder - positively or negatively. Beckman and Fernell (2004) state that children with ADHD in school often have short patience, easily loose their temper and 'explode', interrupt their peers in play and school work and have difficulties to wait for their turn, which often results in school problems. They are often involved in conflicts with peers, have difficulties to handle their feelings, to express themselves and often feel frustrated.

The symptoms vary over time and in different environments although the difference between each individual is huge. Functions related to special thinking processes are limited and they have often limitations in the working memory meaning that they have problems to store and process information and impressions. Many difficulties are thus related to the social environment. The mixture of medical, social and educational factors entail that in order to be successful intervention needs to start with observations of the child in his or her natural context.

This chapter is based on the research project BASTA (Basic skills, social interaction and training of the working memory) which was a longitudinal project with focus on children between nine and twelve years with ADHD or similar symptoms. Only 19 out of 41 children involved in the study have been diagnosed with the tools mentioned above, i.e. DSM-IV or ICD-10. Similar symptoms mean that the undiagnosed children were hyperactive, inattentive, bully, disturbing, had behaviour problems, difficulties to accept social rules and were often involved in conflicts. A majority of them had been placed in remedial classes after recommendation of the teacher without asking the child about his or her opinion. This is also an ethical dilemma. Ljusberg (2009) is very critical to these placements and states that the best of the child is not considered. The parents were often relieved as they could now feel that their child was now placed in a class with more help. This does not promote legal rights of the child, and it conflicts with the UN Convention on the Rights of the Child (1989), which Sweden has ratified.

This chapter will focus on the implications of ADHD in children's everyday life and give some ideas of what actions can be taken to facilitate their daily lives and diminish their difficulties. Focus will also be on their social interaction with peers and teachers in school, as children of this age spend most of their daily time in school and with peers. Education is a right for all children and in the following sections I will introduce the Swedish school system and the growing remedial education.

## 2. The Swedish school system

The Swedish Education Act is based on the UN Convention on the Rights of the Child (1989) and stress education and school for all children. Regardless of gender, ethnic background and social or economic factors all children have equal rights to education on equal conditions in a school for all. Another aspect influencing the Swedish school system is the integration endeavours that can be seen as an obvious result of the Normalization Principle (Nirje, 2003) during the 50´ies and 60´ies. Sweden has gradually moved from a differentiation perspective to an integration perspective, i.e. from one that isolates to one that includes (Brodin & Lindstrand, 2007). Also the Salamanca Statement (2001) has enhanced inclusion and an inclusive school. As a result of this most children in Sweden are today attending the municipal school in their immediate environment often located close to their homes. School is free of charge and compulsory for all children between seven and sixteen years of age. Inclusion is a goal and key concept in the Swedish society and thus also in school. This means that all children are expected to attend the regular schools. No rule without exceptions!

The special schools have closed down and a majority of all children in need of special support today attend the regular school but still there are special programmes for instance for children with intellectual disabilities. However, the number of children with neuro-psychiatric diagnoses and in need of special support in the Swedish schools is increasing and results in an increasing number of special solutions (Matson, 2007). This means that some children still are excluded and segregated in the Swedish inclusive school. These children are placed in small groups or classes, special classes or remedial classes, depending on what expression you choose to use. I will use remedial classes although the meaning is the same - segregation.

### 2.1 Remedial classes

In the municipality of Stockholm the remedial classes increased from 112 in 1998/99 to 180 seven years later (Ljusberg, 2009), and this is contradictory to school policy and research as few studies support exclusion and no support has been found for the superiority of remedial classes (e.g., Karlsson, 2008; Skidmore, 2004). In Sweden, a diagnosis is not needed for placement in a remedial class which means that the main reason for placement often is behaviour problems and the suggestion to move the child to a remedial class often comes from the teacher. Critical voices are raised against this solution and state that some teachers want to have these children moved in order to facilitate their own work. They have not listened to the child's voice and asked for his or her opinion which is against the UN child convention. Previously the diagnosis was closely connected to economic means to the school from the state, in other words 'no money - no support'. This change is a positive development as support can nowadays be given also to children in need of support but without a diagnosis.

Remedial classes have been a reality in Sweden since the public elementary school was introduced in 1842 (Brodin & Lindstrand, 2007). The distinction between regular and special education was earlier that the difficulties should be diminished and that the placement in a remedial class should be temporarily. All pupils in need of special support are, based on their needs, entitled to receive support in school, primarily in the class they normally attend. However, this has turned out to be very difficult and a majority of them are on the opposite moved to a remedial class. Few pupils return to their regular classes. The arguments supporting the establishment of remedial classes have in the past century pointed out two main ideas. The first one is to protect the children and support their self-esteem by shielding them from pupils who perform better (ibid.). Another argument concerns the teacher as these children need much time and space and thus have less time to spend on the other pupils in the class. Children with ADHD in remedial classes may develop a negative identity (Ljusberg, 2009) which means 'I'm a child in need of support, I lack competence and I'm not good enough' and if they are labelled and regarded to be children at risk when they start school, this view will follow them during their entire school careers (Alin Åkerman, 2008; Brodin, 2008).

The remedial class has high teacher density and the average number of pupils per adult and class varies between two and three. Generally, the number of pupils in the class is five to eight. The instructions are given face-to-face and individually to each child. The self-concept of children with ADHD in remedial classes is influenced by the fact that they as a matter of fact are placed in the remedial class. They know that they have got this placement due to their difficulties and failures in school and that they require special support (Taube et al., 1984). Early school experiences will create a feeling of confidence that will help them to proactively enhance their academic development (Brodin, 1997). One problem is that many pupils receive negative feedback on socially unapproved behaviours from parents and teachers (e.g., Teeter, 2004).

> *Negative academic experiences may jeopardize the children's immediate and future views of their academic abilities. At the least; the early negative experience may create a sense of doubt that will likely hinder the student's academic performance and motivation* (Pisecco *et al*, 2001, p 458).

It is reasonable to believe that negative feedback is better than no feedback at all, as the most important for all pupils is to be seen, i.e. to be somebody ('look at me - I'm here'). They often do what they think is the best way in a special context. The self-concept is created in the individual in a social context and as stressed above it is based on biological aspects in relation with the social environment.

## 3. The BASTA project

The BASTA (Basic skills, social interaction and training of the working memory) project was an interdisciplinary project with the overall purpose to explore how training of the working memory with the computer-based programme RoboMemo affected children's basic academic skills and social interaction. The project was based on collaboration between three scientific teams from the medical and pedagogical fields (special education and child studies). The computer-based programme was developed by a research group at Astrid Lindgren's Children's Hospital at the Karolinska Institute (KI) in Stockholm. The responsibility of my team was to focus on child and youth studies, which means that we have the child in the centre and as far as possible try to see things from the perspective of the child in a context.

The main project idea was to get a holistic view of the children involved based on the different perspectives. The sub-studies will be presented at the end of this section, but I will start to give some facts on the working memory as it is central for children's learning and consequently also affects teaching.

### 3.1 Working memory

Working memory research is based to a great extent on neurological findings focused on biological phenomena and presumed genetic defects. The working memory is of great importance for learning and problem-solving and research has shown that it is possible to train the capacity of the working memory (Klingberg *et al.*, 2005; Westerberg, 2004). The working memory helps us for instance to remember what we are going to do next which is necessary in school work. The working memory is used when the child is reading and writing, for mathematics, and problem-solving, and this affects their academic achievements. The programme consists of nine different exercises each designed to challenge and improve the working memory capacity from different angles that is both verbal and visuo-spatial. The degree of difficulty is adjusted and will increase continuously in accordance with the child's achievements as the training progresses. The children (N=41) participated in the training daily during five weeks. Each session was between 30 and 45 minutes and each child involved had an adult (teacher or assistant) aside who was responsible for the training. Criteria for inclusion were average intelligence, deficits in attention with or without the diagnosis ADHD and attending a remedial class in a regular school. Criteria for exclusion were autism, hearing or visual impairments. The children involved were between nine and twelve years of age. Before the children were selected to participate they were interviewed by a psychologist from the project team in order to find out if the child should be included or excluded from involvement in the project.

The interactive computer-based training programme RoboMemo was based on the assumption that the working memory can be improved by training. Research had as mentioned previously showed that significant improvements had been achieved with regard to problem-solving, attention span and impulse control when children with ADHD had been working with the programme (Klingberg *et al.*, 2005). An evaluation of the RoboMemo training programme was conducted in a child habilitation centre in the south of Sweden (Landin, 2007) and the positive results were supported, although the number of participants was very small. Only nine children with ADHD participated in the evaluation. Psychological tests and rating scales showed positive effects. The involved children had improved their working memory capacity and diminished the dysfunction in behaviour but the limited number of participants made it impossible to generalize.

The two research teams at Stockholm Institute of Education (teacher training university) were responsible for child studies and special education. Data have been collected by interviews, questionnaires and classroom observations with children, parents and teachers. Data from 41 children (9-12 yrs old), their parents and 21 teachers in twelve different classes in nine schools have been collected. The special education team conducted studies concerning the basic academic skills in order to find out if and how the training of the working memory affected their skills. They used rating scales and tests and found positive correlations between the training and the performance. The final results of these studies have not yet been published. Four sub-studies were conducted within the field of child and youth studies: self-concept, teaching, ethical issues and children's views. The main reason was to find out if and how the children's self-concept changed after having trained the working memory. Did the improved skills affect their self-concept materially? As we know that these children experience a difficult school day the next focus was on the teacher and the school setting. How is the situation in school for children with ADHD from teacher perspective? What do the parents think about the school situation of their child? Early we noticed that many decisions had been taken 'over the head' of the child and without listening to the child's voice and this resulted in the third study on ethical issues. The forth sub-study focused on how these children thought about their everyday life with special focus on school, which is the arena children attend daily. Together all four sub-studies give a contribution to a greater understanding of children with ADHD and how they think, and hopefully this may diminish school failures. How can peers and adults in the immediate environment support these children in order to increase their well-being? The results of these four sub-studies will be included under the headings below. The concepts children and pupils are used synonymously in the text.

## 4. Self-concept in children with ADHD

Children's creation of identity is based on interaction between what they think about themselves as human beings and on the image they receive from the context, i.e. what other persons think about them (Brodin, 1997). The self-concept can be negative or positive depending on the social environment and therefore this study has a socio-cultural approach. Research on self-concept in children with ADHD is limited and two different lines are visible. Some researchers argue that these children have a low self-concept but they report a higher self-concept than expected (e.g., Hoza *et al*, 2002). Other researchers argue that they have low self-concept and poor self-esteem (e.g., Pisecco *et al*, 2001). Generally, research concerning self-concept in children with disabilities show that most children who experience that they have difficulties have a low self-concept. Children with intellectual disabilities for instance often say they can manage to do things they are unable to and this results in failures and disappointment. In the long run this might influence their willingness to try new things e.g. food, activities etc. This will lower their self-concept and they will most suddenly change their attitude and say 'I can't do anything'. The self-concept is of importance in school work as the pupils must trust their ability to manage.

The purpose of this sub-study was to contribute to knowledge about children with ADHD and how these children rated their self-concept. Most studies focus on children's views from parental and teacher perspectives, but in this study the children themselves are in the centre and the informants (Ljusberg & Brodin, 2007). As appears from research many social, emotional and educational difficulties result in failures in school. The lack of social competence is a huge problem and these children often find themselves in situations they cannot control. The problem occurs immediately and they loose their temper. For their school mates these situations are challenging and many take advantage of it and trig the child with ADHD.

The main interest in this study was to explore if and how training of the working memory affected the children's self-concept and the purpose was to search for knowledge useful for teachers and parents in order to assist these children in school. The data collection involved interviews with parents, researchers and pupils, observations in school and questionnaires. The pupils answered a questionnaire 'This is me' (Taube *et al.*, 1984) and it was used to assess the self-concept from four perspectives: academic, social, personal and global self-concept. Four alternatives to reply were included: YES, yes, NO and no and these were related to the degree of agreement or disagreement. Below are examples of statements in the questionnaire:

The academic self-concept:

I think it is easy to read
I think it is difficult to spell
I need more help from my teacher
I wish I could stop attend school

The social self-concept

They blame me when it is bully in the classroom
I often play by myself on breaks
The other pupils regard me as stupid

The personal self-concept

I would rather be another child
I like to talk to adults
My parents think I'm OK

The global self-concept included all the above-mentioned statements (Ljusberg & Brodin, 2007).

The children completed the questionnaire on three occasions - before the training with RoboMemo started, directly after finishing the training period and finally six months after the working memory training. The questionnaire had been tested on a representative sample of all school beginners (N=690) in 44 regular schools in a town in Sweden and found valid and reliable for the target group (Taube *et al.*, 1984).

The questionnaire 'This is me' was completed in the classroom by the pupils or with assistance of the teacher or assistant, and that is why the result may differ. This was a limitation in the study. The questionnaire had been used by Taube *et al.* previously and allowed us o compare the mean value (M) on all three scales as well as the global scale. Taube reported one occasion of measurement while the BASTA involved three (Ljusberg & Brodin, 2007). However, we do not have access to the standard deviances in the comparison data. The dominating pattern was that the differences were small to nonexistent between the BASTA data and the comparison data concerning global, academic, social and personal self-concept. Few changes are visible in the material with regard to the self-concept of the children. We know however that RoboMemo improved the working memory capacity and this resulted in a higher capacity of the skills in reading, writing and mathematics. The increased academic skills might have affected the self-concept in the long run as a higher intellectual skill often influences the self-confidence and the self-esteem. Remarkable was that the children reported a higher degree of self-concept than expected although most people think that entering a remedial class is in a way linked to academic and social failures.

The main question is to find out what these children learn in school as the main task in school is learning in a wide perspective. The children said they had a good self-concept but it appears from the questionnaire that they overestimated their self-concept. That they reported a higher self-concept than expected may confirm the theory of protection. Some researchers mean that the inflated self-concept serve a self-protective role for children with ADHD despite experiences of failure. Awareness of this may facilitate for the teachers. In other words: Boys with ADHD can relax their self-protective position once they know it is not needed and from a pedagogical point of view this is an interesting finding. As Ljusberg (2009) expressed it: 'they can lower their guards'. The main result was that despite repeated failures children showed a fairly good self-concept. However, they still need positive support from persons in the immediate environment to succeed in school work and in school the key persons are the teachers. In the next section teacher education and teaching will thus be in focus.

## 5. Teaching children with ADHD

Teaching in today's school is demanding and challenging and many teachers state that they experience difficult teaching situations with too many pupils in the class and a shortage of special education knowledge. An observation is that many teachers leave their jobs shortly after having finished their teacher training as the job was not what they expected it too be. It is too demanding and they state that they don't feel that they have knowledge enough to work with children in need of special support. The weak connection between theory and practice is especially criticized and the shortage of special solutions to classroom problems combined with the fact that children in many schools have complex needs and mental health problems are realities that have not been stressed strongly enough in teacher training (Brodin, 2011). Another aspect is that inclusion demands both careful teacher training and time to make changes (Brodin & Lindstrand, 2007). Inclusion is expected to give equal opportunities and participation for all children but the number of remedial classes is ever increasing. This is especially true for children with neuro-psychiatric disorders, who often have difficulties to take part in education in a regular class.

For children who have ADHD the school day is often problematic as they lack social competence (Barkley, 2006). They are often involved in conflicts with peers and teachers and they are regarded as bully and disturbing. Their difficulties to concentrate often result in inattentiveness and when the teachers give instructions to the class their thoughts are often somewhere else. They miss a lot of information and when other pupils listen to the instruction and start to work on the task these children have lost the whole message (Brodin & Ljusberg, 2008; Hellström, 2010). They have not understood the task and they cannot catch up as they have not 'heard' the instructions. Besides it is impossible for these pupils to remember long instructions from the teacher as they have limitations in the working memory capacity. This is why it is important to train the capacity of the working memory. The main problem for these children is that they feel stupid and are anxious to ask the teacher to repeat the instruction again. Instead they choose to skip the task without giving any reason, which of course is the easiest way to get away from the tiring situation. If the teacher divides the instructions in smaller sequences it would be a great help to these children. The teacher must realize that these children have many difficulties partly due to the large number of failures they have earlier experienced in school.

The interviewed children were fully aware why they attended the remedial class. One problem in school is that children with ADHD influence their classmates' opportunities to concentrate and learn due to their misbehaviour and they hinder the teachers to teach effectively. The rules recommended by Gillberg (2005) i.e. small teaching groups or remedial classes, individual instructions in reading, writing and mathematics, short sequences for concentration and a number of physical activities, are still very useful. Structure and routines will benefit all children. Another aspect is that the tasks in school must be of interest for the pupils and it is for many teachers challenging to find tasks for all.

The purpose of this study was to increase the teachers' and parents' knowledge of the school situation for children with ADHD and data were collected by a questionnaire to the teachers (N=21) and parents (N=40). One parent dropped out. The classes were mixed with pupils of different ages and 24 pupils had attended the class for less than three semesters. A majority of the children participated in activities organized together with the pupils from the regular classes, which means that the school and teachers have tried to meet the goals of inclusion. The teachers mentioned that short breaks, gymnastics, lunch breaks and outdoor education were the most common activities they spent together. Many pupils with ADHD or similar symptoms had inferior skills and produced academic results that left them at the bottom of the class.

Most teachers worked full time in the remedial class and half of them had been working in the class between one and three semesters, which is a short time. This is definitely not benefiting these pupils. It was evident that these children had to change teachers often. Many teachers were rather old and had an old basic teacher training and that is a well-known fact from other European countries. Old teachers do not necessarily mean that they are poor teachers but they have an old education and when they were in teacher training few talked about children with neuro-psychiatric disorders. Another problem is that many schools also had difficulties to recruit teachers. A majority of the parents (N=34) expressed that they were satisfied with the school situation of their child and they did not have any complaints about it. Some were also satisfied that their child had been moved to the remedial class as they believed that the child would now get all the support he or she needed. The parents said that they were relieved when the disorder of the child was confirmed.

This reminds me on the situation for children with intellectual disabilities in Sweden long time ago. When they were placed in institutions the parents expected them to get the very best education and help but many years later they realized that their children had just been placed at an institution without the expected support.

A majority of the teachers in the BASTA project had received their teacher education during the 1960s and 1970s. None had experience of special education. Seven of them had taken courses and got in-service training and they regarded themselves as competent for the job and added that 'I get enough experience in the classroom'. The teachers experienced a tough time working with these pupils and only half of them stated that they got supervision. The parents reported that they relied completely on the teachers' skills. Most of the teachers had extra staff in the classroom to support the pupils and thus felt satisfied. The parents were sure that their son or daughter had good support. Some parents however pointed out that educated young teachers ought to be recruited, as they would have more energy to work with these restless and trying children. With regard to the staff turnover and stability in the class both teachers and parents seemed to be satisfied and the teachers pointed out that they found the classroom climate better in the remedial class than in the regular classes. The teachers said that 'for these children the remedial classes are the best solution'.

The teachers highlighted the difficulties they had to teach in the remedial class as the pupils are disorderly, restless, hyperactive and inattentive. Many of the pupils could only concentrate a couple of minutes at each time and then started to move around in the classroom. Much time was spent on correction of the pupils' unacceptable behaviour, and to make them listen to their instructions instead of teaching.

Despite that these children had been placed behind screens and work for themselves without distraction and interruption from others, they disturbed each other by throwing paper balls or rubbers at their peers. These disturbances of course also existed in regular classes and in other educational contexts. As their span of attention is extremely low they are distracted by other activities going on and easily lose the red thread. Barkley (2002, 2006) states that these children as soon as they get an opportunity do what they found most tempting for the moment. The situation would probably not be better in a regular class.

The question is why parents are so satisfied with the placement in the remedial class. At least two explanations are visible. First of all they believe that their child will get a lot of help and they do not need to worry about the child's education, secondly that they avoid to get complaints about their child's behaviour from the school as the child's behaviour is more or less accepted in the remedial class. Another explanation is that the parents realize that despite the teachers' lack of special education, complaints will not help. When the child is placed in a remedial class the parents can relax instead of worrying about the behaviour of their child. Also the teachers and school staff seem to be satisfied with the situation which seems strange. How can they feel satisfied when they cannot teach as they want to? Why don't they ask the headmaster for more resources if they cannot do their job in a way good enough?

We know that the great number of academic and social failures has contributed to a low self-concept and confidence in children with neuro-psychiatric disorders. But how are these pupils treated in school? What are they expected to learn in school and what are they actually learning? Ljusberg (2009) was very provocative when she said that what these children primarily learn 'is to be pupils in a remedial class' (p. 51). The teachers often stressed that they train the children to behave nicely in order to be transferred to a regular class in the future - but how often does this happen? Few pupils have been and will go back to the regular class and in the meantime they loose a lot of opportunities to learn. 'A relevant question is if these children learn more in a remedial class than in a regular class, or if the small class is chosen as en emergency solution for the teachers only?' (Brodin & Ljusberg, 2008, p. 354). Evidently the pupil is bearer of the school's problem (Karlsson, 2008).

## 6. Interviewing children with ADHD and ethical aspects

Children constitute an overexposed group in research because they are under age and because they are in a dependent relationship with adults. What can we learn from interviews with children with ADHD in remedial classes - what do they want to tell us? This sub-study was conducted with ten children between ten and twelve years of age, who were attending ten different remedial classes. The aim was to stress how children with neuro-psychiatric disorder described themselves and why they were attending the class, as most of these children actually wanted to attend a regular class. Semi-structured interviews formed the basis and the results showed that the children 'are carriers of their schools compensatory perspective' (Ljusberg, 2010). These children know that they are regarded as difficult, annoying and problematic and they know that they are moved to the remedial class because they were so problematic in the regular class that the teacher could not manage. They are fully aware of their situation and confirm that they have difficulties, are restless and inattentive and they blame themselves. These young children are between ten and twelve years old and they take over the whole responsibility for the placement. Ljusberg (2009) states that it is the remedial class per se that creates social difficulties for these children. When working with the interview study we noticed that there are many ethical issues coming up which we had not regarded carefully enough (Ljusberg, Brodin & Lindstrand, 2007). The main reason for highlighting ethical issues in child and childhood research is to protect the involved children and to avoid negative consequences from participation in the project. Ethical issues in child studies are an area that must be stressed more often and in a broader perspective.

Lately, research concerning children's perspectives has increased. Children are under age, they are used that adults talk about them and over their heads and often their opinions have minor influence (UN, 1989). Normally the child possesses information that the researcher wants, i.e. the child's thoughts and opinions, and the ambition was to listen to their voices. One ethical issue is why we build up a school with remedial classes and segregate children who are not good enough. These children are often regarded as troublemakers and few evaluations have been done in remedial classes (Ljungberg, 2001; Teeter, 2004). Earlier interventions on children with ADHD in classroom settings have shown minor improvements in academic skills and behaviour (Tannock & Martinussen, 2001), so we cannot argue that we make life easier for these children by placing them in a special class. The ethical issue concerns how we can defend this conscious exclusion when we are talking about inclusion and a school for all. Two controversy pictures of reality.

Before children are involved in research a written permission of the parents is needed but the child also needs to accept the conditions for being interviewed. When children are involved it is also important that there are no bonds or relationships of dependency between the researcher and the child. The risk of identification is always a problem when working with children in difficult situations and we therefore tried to hide data that could be identified by making summaries. Some of the children felt special as they were selected to participate and sometimes gave information that they would probably not do otherwise, and this might be a dilemma. Being a child in a remedial class implies exclusion from ordinary teaching.

> *One boy (10 years old) told us about his special situation and the anger and distress he felt. He blamed his mother for being placed in the remedial class and also mentioned other things that irritated him about the mother. Suddenly he said that he should write a letter to his mother and tell her that he had moved from home. The interviewer then said that the best he could do was to tell his mother that he was angry with her. The boy replied that he felt that his father cared more about him than his mother* (Ljusberg, *et al.*, 2007, p. 207).

The conversation went on and the boy had used the interviewer to test his ideas. After that the interview continued as normal. This boy really expressed his dissatisfaction with being a pupil in a remedial class but he had problems to talk about it with his parents. Sometimes children give sensitive information; later they might feel that they have gossiped about someone. The researcher often receives confidential information and must take responsibility for how it is used, but without being a therapist. The most important is to listen to children's voices and to create learning environments in school where all children feel welcome and good enough from their prerequisites (Brodin, 2011; Danby & Farrell, 2004). Equal opportunities and participation is an ethical issue.

## 7. The school setting

To summarize, the difficulties for children with ADHD in everyday life primarily concern domains such as education, social relations, self-esteem, self-concept, family functioning and independence. The school is an arena attended by all children and learning is the most important task in school (Brodin & Lindstrand, 2007; Brodin & Ljusberg, 2007). The function we use to store information is the working memory and this function is central for attention and concentration on a task but also to control impulses. The working memory is thus vital for learning and problem-solving as it helps us to remember what we are going to do next which is important in school work. It works when the child is reading and writing, for mathematics and these skills affect their academic achievements. Ljungberg (2001) suggested that ADHD can develop differently depending on the social environment and due to this the school setting is of great importance. Based on reported research I support the combination of medical, social and educational factors.

In today's society we are flooded with information and to handle this information we need to coordinate and interpret what we see and hear. Our brain needs to be able to judge whether the information is valuable or not. Many brain functions can be compared to computerized calculations, e.g. memory, selection of information and working memory. The function of the working memory is to keep information actual in the brain and this function is developed during the first years of life. Memories of facts and actions will be stored in the working memory while they are coded into the long term memory. For children with neuro-psychiatric disorders training of the working memory capacity is decisive for their learning (Westberg, 2004).

Our observations in the classroom showed that the pupils commonly were placed one by one with the desk placed against a wall. Between the pupils a screen was placed in order to avoid distraction of other children. The main reason for this was to do the best for the children as they are inattentive and easily loose their interest for a task. There are small talk going on between the teacher and the pupils and the climate in the classroom seemed to be calm and friendly. The teachers' ideas are to compensate for the disorder and they explain that their role is to meet every child where he/she is (Brodin & Ljusberg, 2008; Teeter, 2004). The teachers in the remedial class talk in a critical way about the teachers working in the regular school. They also state that pupils with ADHD have problems the school cannot solve. When looking at the so called classroom – it is easy to understand that visitors in the class must question the placement of the pupils. The organization of the classroom is of great importance and it shows obvious what expectations the teachers have on their pupils. Hellström (2004) talks about 'self-fulfilling prophecy' and many of these children follow the pattern expected from them. One of the main questions is how interaction between the classmates works. They say that they have minor social exchange outside school and at breaks during the school day. Is a hard structure of the classroom a good solution from an educational point of view or can the school setting better be adapted to fit all children in the class? Probably not! The strict structure in the classroom just confirms and conserves the deviant behaviour in the school setting. How much of individual teaching is needed to stimulate their interests for learning? The main task for the school must be education and to give pupils positive educational and relational experiences that will make life easier for them.

## 8. Social and educational implications of ADHD in school

In the western world we can today note an increase of children with difficulties in managing and adapting in the regular school and many children with neuro-psychiatric disorders will in the long run drop out of school. This situation is not acceptable from the child's perspective. They have the right to learn and they must be given equal opportunities in the same way as other children of their age. These children cause a huge problem for the teachers who often lack knowledge of special education related to this disorder (Brodin & Ljusberg, 2008) and based on this Teeter (2004) recommends that intervention based on reliable tests should be done. In many countries children with ADHD are transferred to remedial classes and segregated from their classmates in regular classes and many researchers are sceptical to this solution (e.g., Ljusberg, 2009). They state that many pupils presumed at risk when they start school are seen in the same way during their entire school careers which means that they are labelled for life. This is also what many parents are afraid of. Some parents mean that ADHD is a negative label and they sometimes reject to see the difficulties in the child as they do not want to have their child labelled. But this is a mistake as the child cannot get the help he/she needs. The question is when 'the best of the child' is stated in the UN Convention (1989) - what does it really mean? Whose interest is in focus? The parents might feel responsible for the child's behaviour but hope that the complicated situations and difficulties will disappear when the child gets older. In the meantime these children feel misunderstood and that they do not fit in. The well-being of the child is set aside (Brodin, 2011). The choice for these children is to be segregated and placed in a remedial class for children with ADHD which is located in a regular school. The placement in remedial classes has negative consequences with regard to inclusion and many researchers (e.g., Karlsson, 2008; Skidmore, 2004) argue that there is no support for the superiority of remedial classes. But in Sweden the number of special classes for children with different disorders is increasing. The discrepancy between political goals (an inclusive school) and reality (special solutions) is a fact. At the same time we get reports that many school-age children feel misunderstood are unhappy and do not feel well in the Swedish school.

## 9. Communication and social interaction

Although the main task in school is learning basic academic skills such us reading, writing and mathematics also other goals e.g. communication and social interaction are vital. Children with ADHD often have difficulties in communication and social interaction and many children experience a problematic school day as they lack social competence to interact with their schoolmates and their teachers (Barkley, 2006; DuPaul & White, 2006; Kadesjö, 2010). This mainly depends on their difficulties in listening to people around, e.g. to the teacher's instructions and to finish a task in school (Westerberg, 2004). They get easily involved in conflicts with peers as they have difficulties to adapt to approved social rules and to control their impulses. In order to find a way out, they get involved in a fight. Children who cannot express themselves use the fists to convince the peers. As a whole these children are regarded as bully and disturbing by peers and teachers (Brodin & Ljusberg, 2008), and interviews with these children showed that they blame themselves for being placed in the remedial class. 'It is my fault - I can't behave - I'm bully'. They often have failed to build up relations with peers both inside and outside of school and many of these children feel very lonely since they have been moved to the remedial class. They have lost many of their friends and they have difficulties in building up new relations. In a way they have a realistic self-concept because they repeat all the time that they are problematic and that the other schoolmates tease them and call them 'DAMP-children'. There are many negative aspects with ADHD but there are also positive aspects. Many of these children are seen as exciting and challenging as they dare to break rules and follow their instincts. A lot of exciting but sometimes dangerous things happen around them. Their thinking is often unconventional and they are extremely brave to try new activities as they do not consider the consequences. They are also extremely concentrated on things they really found interesting. Often we hear about children who have learnt things that have no real meaning for most of us. The challenging issue for the teacher is to find out exactly what interests each child has. All children are unique and differences between children are challenging but stimulating. If they all were the same nothing exciting would happen in life.

## 10. What can we do to help these children to lead a good life?

Children with ADHD have equal rights to education as all children. Most of the research today highlights the problems with these children and this article also confirms that there are huge problems in school. Most countries are working towards an inclusive school, but in spite of that most countries segregate these children and place them in special classes. The reason is that they are too bully and disturbing for their classmates and for the teacher. The teachers often complain about the difficult classroom situation and the lack of moments of 'possible teaching'. The main question is what we can do to facilitate education and learning for these children?

Many aspects influence learning for instance the teacher training, the teacher's view of the child, the parent's acceptance of the child's disability, the learning material in school, the school tasks, and the school setting including the structure of the classroom.

The *teacher training* is crucial. The teachers need more knowledge of special education in their regular teacher training. They must be trained to work with all children, despite difficulties and prerequisites and they also must be trained to meet the needs of children with different backgrounds and disabilities with respect. This is partly an educational aspect, partly a human aspect i.e. how they view the children with ADHD, what prejudices they carry. Hellström (2004) talks about full-filling prophesy and it is dangerous if the teacher already from the start have an idea of the functioning of each child. To label children for life creates a problem - they need a second chance. If the school supported all children based on their special needs this would probably be the best for each child (typical or atypical). Flexibility is a word that should be highlighted more often. Structure and routines are also useful concepts to support all children. Consequently, teachers need a better basic training and they need at least a little knowledge about how the brain develops in a young child in order to have realistic expectations on the child's learning opportunities. The tasks must be of interest for the child and this is a major problem also in the regular school. The school is sometimes very old-fashioned and the norms of the school are difficult to understand for today's children. We live in the digital age and many teachers are not updated on how much influence ICT has on children's world. Children are always children of his or her time.

Today there are also a large number of programmes for behaviour modification available on the market. Teachers need training in how these programmes can be used to change the behaviour of the child. The main principle of modification is to reinforce and praise all activities and behaviour that is approved and nice and to neglect misbehaviour. Some schools are working with token economy which means that a good behaviour renders some kind of reward. This will often stimulate the child to modify his or her bad behaviour.

Also the *parents* need knowledge of ADHD and what implications it has on the child's daily life. In many countries special parent education is organized where the parents of a child with disorder can meet other parents with the same interests (Chandler, 2010). Intervention must always involve the child's entire situation both at home and at school and the methods used for treatment should be evidence-based in order to be successful. Normally the treatment starts with changes in the immediate environment, i.e. to adapt the school setting to fit the needs of the child. Sometimes this is not enough and medication is used as a complement. Intervention and treatment of children with ADHD in school are constantly on the agenda and a highly discussed topic in media. A small group of medical doctors reject to prescribe Ritalin or similar medication to children while most neuro-psychiatric doctors see it as a complement to environmental aspects (Eriksson & Ingvar, 2004).

As mentioned above special training programmes might help the child to perform better. One example is the RoboMemo programme for training of the working memory capacity

(Klingberg, *et al.*, 2005) that has been of great help for many children concerning basic academic skills. In the long run it is reasonable to believe that a child who performs better in school also gets a changed and stronger self-concept. There are also a quite large number of pedagogical methods used today and the development of new ways to treat these children is constantly on its way. The tasks in school and the material used must sometimes be changed and developed and the teachers must often use their fantasy to find solutions. Teachers and parents must cooperate in order to strengthen the confidence in these children, and structure and routines will contribute to their well-being (Brodin, 2011).

Finally, the school setting is of importance. The classroom is structured to help the child, but if the remedial class is only an occasional placement, then it will be difficult to go back to the regular class as the immediate environment will look so different. The feeling when coming into a class with no natural social interaction between the pupils is strange. If they want contact they just throw away a rubber or a piece of paper on somebody, who immediately reacts with anger. Perhaps a negative contact is better than no contact at all for these children. As many of these pupils are hyperactive the structure is a way to keep the atmosphere nice and calm in the classroom. The interviews with the children with ADHD showed that 'the school focused on the pupil's shortcomings instead of the pupil in the context' (Ljusberg, 2009, p 47).

Still research from a pedagogical aspect is limited and more research should be conducted in order to better meet the needs of these children. Some aspects seem more important than others; high teacher density, qualified teachers, small classes and few pupils, short and clear instructions, many breaks, adapted tasks, cooperation between parents and teachers, intervention and suitable treatment, e.g. to use evidence-based pedagogical programmes for behaviour modification. Besides, respect for the child, to do the best for the child, to listen to the voice of the child, and to invite the child to have influence on the teaching and whole school setting are important aspects to facilitate daily lives of these children.

## 11. References

Alin åkerman, B. (2008). Psykisk ohälsa och risk för självmordshandlingar bland ungdomar [Mental unhealth and risk for suicide among youngsters].In J. Brodin (Ed.) Barn i utsatta livssituationer [Children in exposed living situations] (pp. 133-170). Malmö: Gleerup. Education.

Brodin, J. (1997). Självuppfattning hos personer med intellektuella funktionshinder. [Self-concept in children with intellectual disabilities] TKH-report 15. Stockholm: Stockholm Institute of Education.

Brodin, J. (2008). *Barn med funktionsnedsättningar* [Children with disabilities]. In J. Brodin [Ed.) Barn i utsatta livssituationer [Children in exposed living situations]. Malmö: Gleerup.

Brodin, J. (1997). Självuppfattning hos personer med intellektuella funktionshinder. [Self-concept in children with intellectual disabilities] TKH-report 15. Stockholm: Stockholm Institute of Education.

Brodin, J. (2009). *Support systems in preschool and school for children with disabilities*, IRIS (Improvement through Research in the Inclusive School), EU project. www.irisproject.eu

Brodin, J. (2011). Children in precarious environments and life situations. *International Journal of Child and Adolescent Health* 4(2), 131-138.

Brodin, J.& Lindstrand, P. (2007). Perspectives of a school for all. *International Journal of Inclusive Education*, 11(2), 133-145.
Brodin, J. & Ljusberg, A-L. (2008). Teaching children with attention deficit hyperactivity disorder in remedial classes. *International Journal of Rehabilitation Research*, 31(4), 351-355.
Barkley, R. A. (1997). *ADHD and the nature of self-control.* New York: Guilford.
Barkley, R. A. (2002) (Ed.). *International Consensus Report on ADHD.* University of Massachusetts Medical School: Dept. of Psychiatry and Neurology.
Barkley, R. A. (2006). *Attention Deficit - Hyperactivity Disorder: A handbook for diagnosis and treatment* (3rd ed.) New York: Guilford.
Beckman, V. & Fernell, E. (2004). Utredning och diagnostik [Intervention and diagnostic]. In v. Beckman (Ed.) *ADHD/DAMP - en uppdatering* [ADHD/DAMP - updated]. (pp 21-35). Lund: Studentlitteratur.
Chandler, C. (2010). *The science of ADHD: A guide for parents and professionals.* Malden, MA: Wiley-Blackwell.
Danby, S. & Farrell, A. (2004). Accounting for young children's competence in educational research: new perspectives on research ethics. *The Australian Educational Researcher*, 31(4), 35-49.
Diagnostic and Statistical Manual of Mental Disorders DSM-IV, rev. 2002. NY: American Psychological Association (APA).
DuPaul, G. J. & White, G. P. (2006). ADHD: Behavioral, Educational and Medication Interventions. *Education Digest*, 71(3), 57-60.
Eriksson, E. & Ingvar, M. (2004). About ADHD: an actual fight going on in Sweden. In v. Beckman (Ed.) *ADHD/DAMP – en uppdatering* [ADHD/DAMP - updated]. (pp 123-132). Lund: Studentlitteratur.
Gillberg, C. (2005) (2nd. ed.). *Ett barn i varje klass. Om DAMP, MBD, ADHD* [One child in each class. On DAMP, MBD, ADHD]. Stockholm: Cura.
Harwood, V. (2006). *Diagnosing 'Disorderly' Children. A critique of behaviour disorder discourses.* London: Routledge.
Hellström, A. (2004). Psykosociala och pedagogiska stödinsatser [Psychosocial and pedagogical interventions. In. V. Beckman (Ed.) *ADHD/DAMP – en uppdatering* [ADHD/DAMP - updated]. (pp 101-122). Lund: Studentlitteratur.
Hellström, A. (2010). *Att vara förälder till ett barn med ADHD – så kan du underlätta vardagen* [To be parents of a child with ADHD - how to facilitate everyday life]. Stockholm: ADHD-centre.
Hoza, B., Pelham, WE. Jr., Dobbs, J., Owens, J., & Pillow DR. (2002). Do boys with attention-deficit/hyperactivity disorder have positive illusory self-concepts? *Journal of Abnorm Psychology* 111, 268-278.
Ingvar, M. (2004). *Uppmärksamheten och hjärnan* [Attention and the brain]. In V. Beckman (Ed.) ADHD/DAMP - en uppdatering [ADHD/DAMP - updating] (pp 35-47). Lund: Studentlitteratur.
Kadesjö, B. (2008). *Barn med koncentrationssvårigheter* [Children with concentration difficulties]. Stockholm: Liber.
Kadesjö, B. (2010). *Barn som utmanar – Barn med ADHD och andra beteendeproblem* [Challenging children - Children with ADHD and other behaviour problems]. Stockholm: The National Social Board of Helth and Welfare. Electronic resource.
Karlsson, Y. (2008). *Att inte vilja vara ett problem – social organisering och utvärdering av elever i en särskild undervisningsgrupp* [Resisting problems - social organisation and evaluation practices in a special teaching group]. Linköping: Linköping University. Dissertation.

Klingberg, T., Fernell, E., Olesen, P., Johnson, M., Gustavsson, P., Dahlström, K., Gillberg, C., Forssberg, H. & Westerberg, H. (2005). Computerized training of working memory in children with Attention.Deficit/Hyperactivity Disorder - A randomized, controlled trial. *Journal of the American Academy of Child and Adolescent Psychiatry*, 44(2), 177-186.

Landin, I. (2007). RoboMemo - en utvärdering av arbetsminnesträning för barn med ADHD [RoboMemo - an evaluation of working memory training for children with ADHD]. Report 8/2007. Malmö: Dept. of Research and Development, www.skane.se/habilitering/fou

Ljungberg, T. (2001). *ADHD hos barn och ungdomar – Diagnostik, orsaker och farmakologisk behandling.* [ADHD in children and youngsters - Diagnostics, causes and pharmacological treatment]. Stockholm: The Social Board of Health and Welfare.

Ljusberg, A-L. & Brodin, J. (2007). Self-concept in children with attention deficits. *International Journal of Rehabilitation Research*, 30(3), 195-201.

Ljusberg, A-L. (2009). *Pupils in remedial classes.* Stockholm University: Department of Child and Youth studies. Dissertation.

Ljusberg, A-L. (2011). Children's views on attending a remedial class - because of concentration difficulties. *Child: care, health and development.* iFirst 2011, 1-6.

Ljusberg, A-L. & Brodin, J. (2007). Self-concept in children with attention deficits. *International Journal of Rehabilitation Research*, 30(3), 195-201.

Ljusberg, A-L. & Brodin, J. & Lindstrand, P. (2007). Ethical issues when interviewing children in remedial classes. *International Journal of Rehabilitation Research*, 30(3), 203-207.

Lloyd, G. (2006). Conclusion: Supporting children in School. In G. Lloyd, J Stead & D. Cohen (Eds.) *Critical new perspectives on ADHD*, (pp 215-228) London: Routledge Falmer.

Matson, I-L. (2007) *En skola för eller med alla* [A school for or with all children]. Stockholm: Stockholm Institute of Education. Licentiate Dissertation.

Nirje, B. (2003). *Normaliseringsprincipen* [The Normalization Principle]. Lund: Studentlitteratur.

Norwich, B. (2008*). Dilemmas of difference, inclusion and disability. International perspectives and future directions*. London: Routledge, Taylor & Francis group.

Pisecco, S., Wristers, PS., Silva, PA., Baker, DB. (2001). The effect of academic self-concept on ADHD and antisocial behaviours in early adolescence. *Journal of Learning Disability*, 34, 450-461.

Salamanca Statement (2001). Stockholm: The Swedish Board of Unesco.

Socialstyrelsen [The National Board of Health and Welfare] (2004). *Kort om ADHD hos barn och vuxna: en sammanfattning av socialstyrelsens kunskapsöversik*t [Brief on ADHD in children and adults: a summary and literature review]. Stockholm: The National Board of Health and Welfare.

Skidmore, D. (2004). *Inclusion - the dynamic of school development*. Maidenhead: Open University Press.

Tannock, R. & Martinussen, R. (2001). Reconceptualization ADHD: *Educ. Leadership*, 59, 20-26.

Taube, K., Torneus, M. & Lundberg, I. (1984). *Umesol, Självbild* [Umesol - self-concept] Stockholm: Psychology Press.

Teeter, P. A. (2004). *Behandling av ADHD – ett utvecklingspsykologiskt perspektiv* [Treatment of ADHD - a developmental pschychological perspective]. Lund: Studentlitteratur.

UN Convention on the Rights of the Child (1989). New York: United Nations.

Wells, KC. (2004). Treatment of ADHD in children and adolescents. In PM Barrett and TH Olledick (Eds.) *Handbook of interventions that work with children and adolescents: Prevention and treatment*. Chichester, UK: John Wiley. Pp. 343-368.

Westerberg, H. (2005). *Working Memory: Development, Disorders and Training*. Stockholm: Karolinska Institute. Dissertation.

# 9

# Evaluation of the Level of Knowledge of Infant and Primary School Teachers with Respect to the Attention Deficit Hyperactivity Disorder (ADHD): Content Validity of a Newly Created Questionnaire

Marian Soroa, Nekane Balluerka and Arantxa Gorostiaga
*University of the Basque Country*
*Spain*

## 1. Introduction

ADHD is a universal disorder which began to be researched more than a century ago but about which there remain many questions regarding its etiology, evolution and effective treatment (Moreno, 2008b). Without doubt this is, in part, due to the fact that people with ADHD are a heterogeneous group with varied symptoms and not all cases have all the symptoms and features which have been described as characteristic of this disorder. In general terms it can be stated that people with ADHD suffer from an inappropriate development of the mechanisms that regulate attention, reflexivity and activity (Miranda, Jarque & Soriano, 1999).

In spite of the fact that so much remains unknown about the disorder, ADHD is one of the most common psychological problems in childhood and adolescence (Brown, 2003; Barkley, 2004) and produces significant negative consequences for those who suffer from it. As Barkley (1998) pointed out, the importance of ADHD derives not only from its direct effects but also from the contribution it makes to increasing the vulnerability of the person suffering from it to other problems and difficulties. Although it is not possible to construct a unitary profile which would bring together all those affected by the condition because of the heterogeneous nature of the symptoms they suffer from, the majority of children with ADHD suffer from, as well as the central symptoms of the disorder, other associated problems which make the prognosis of the disorder even more difficult. The most common associated problems are usually disruptive behavior and problems at school, cognitive and social difficulties, emotional disorders, physical problems and associated disorders, particularly oppositional defiant disorder, conduct disorder, anxiety disorders and learning disorders (Moreno, 2008b).

Currently ADHD is one of the conditions that generates most research in the scientific community but even so, in the words of Barkley (2005), it continues to be largely unknown and misunderstood. Teachers, along with the family, are one of the most important agents of socialization during infancy so they are one of the groups most suited for training and awareness raising with regard to ADHD. As will be pointed out in the following section, a significant percentage of teachers have false ideas or gaps in their knowledge regarding ADHD, which leads them to act in a mistaken fashion in the classroom. Children with ADHD require more attention than other students, a series of organizational and structural changes, and more involvement by teachers. Thus it has been observed that special training for teachers on how to deal with this issue and a positive predisposition on their part can have very positive results for children with ADHD.

### 1.1 ADHD and teachers

The amount of research into teachers' knowledge of ADHD has grown considerably in recent years due to the important role this group plays in the academic, personal and social success of students with ADHD. According to Pfiffner (1999), the teacher is the decisive element in the success of the child with ADHD.

Basing our view on the existing literature, we believe that infant and primary school teachers should have specific and general knowledge of ADHD for six fundamental reasons. Firstly, it must be borne in mind that ADHD is one of the most common psychological disorders among children. The American Psychiatric Association (APA, 2000) claims that between 3% and 5%, approximately, of schoolchildren have ADHD, that is to say, it is estimated that there is an average of 1 child with ADHD in every classroom of 25 children (Barkley 1999; Moreno & Servera, 2002; Moreno, 2008a). Thus, general education teachers in infant and primary schools may encounter an average of one child with these characteristics per school year while specialist teachers of physical education, foreign languages, music or special needs may encounter more than one child in their classrooms with ADHD per school year.

Secondly, it has been shown that the majority of children with ADHD exhibit behavior significantly different in many respects from that of their peers during their pre-school years (Miranda, Roselló & Soriano, 1998) and that these differences become more evident in the early years of primary education when academic and social demands increase (Parellada, 2009). In both these stages teachers are in a good position to identify possible cases of ADHD. They can pretty accurately distinguish between what is normal development and what is not, which leads to their being the professionals who most commonly make initial referrals for specific evaluations of whether children have ADHD (Jarque, Tárraga & Miranda, 2007; Vereb & DiPerna, 2004). Until recent years pre-school teachers preferred to wait, thinking that the typical symptoms of ADHD were transitory in nature and would disappear with the passage of time. It is fortunate that nowadays more and more of these teachers are aware that ignoring these early indications could constitute a serious error because it could lead to vital time for treatment being lost (Miranda et al., 1998). Thus, increasing teachers' awareness of ADHD could facilitate, among other things, the early detection of the disorder and the provision of appropriate treatment for it (Jarque et al., 2007; Ohan, Cormier, Hepp, Visser & Strain, 2008).

Thirdly, it must be recalled that the role of the teacher is also essential in the establishment of the diagnosis. The teacher's assessment of the behavior of the student, along with that of his or her parents and the results of the rest of the tests administered to the child, forms part of the data that allow the evaluator to establish the diagnosis (Jarque et al., 2007; Sciutto, Terjesen & Bender, 2000; West, Taylor, Houghton & Hudyma, 2005).

Fourthly, it should be noted that teachers play an important role in the implementation, evaluation and support of the treatment of children with ADHD (Ohan et al., 2008). Their support is needed for the results of the treatment received by the student to be successful, and their views and opinions about ADHD have a profound effect on the treatment's efficacy (Sherman, Rasmussen & Baydala, 2008). In this context it has been demonstrated that teachers' knowledge of effective treatment for ADHD has an effect on the support they provide for those treatments (Ohan et al., 2008). In general it has been found that teachers prefer treatments that have positive consequences for students, are easy to implement and require little time (Graczyk et al., 2005). In any case, the research that has been carried out indicates that the knowledge teachers have about the design and implementation of treatment is frequently deficient (Arcia, Frank, Sánchez-LaCay & Fernández, 2000; Sciutto et al., 2000; West et al., 2005).

A fifth reason for teachers to be trained about ADHD arises from the direct contact that they have with the parents of the children concerned. Various authors have found that teachers make recommendations to parents -both correct and mistaken- about ADHD and the parents tend to follow them (Kos, Richdale & Hay, 2006; Ohan et al., 2008). This does not constitute a problem in cases in which the teachers are well informed and well trained regarding the condition, but the advice given by untrained or inadequately trained teachers can be very damaging for the children and their families.

A sixth and final reason arises from the fact that the teacher's knowledge of ADHD has an effect on his or her conduct and attitudes towards the children affected by this condition (Barkley, 2006). Better informed teachers have more positive attitudes and conduct towards students with ADHD (Bekle, 2004; Ghanizadeh, Bahredar & Moeini 2006; Kos et al., 2006). Knowledge of ADHD, which is mainly acquired from specific training focused on the condition and previous exposure to children with it, seems to be associated with the level of confidence teachers have in their own abilities to respond to the needs of their students with ADHD (Bekle, 2004; Graczyk et al., 2005; Jarque et al., 2007; Kos et al., 2006). Bandura (2001) holds that this self-confidence, or self-efficacy is learned and is principally acquired by way of vicarious learning and by direct practice. Furthermore, Graczyk et al. (2005) state that this self-confidence, or self-efficacy is associated with the knowledge the teacher has about the treatments that can be applied to children with ADHD and that educators usually show little confidence in their abilities when it comes to responding to the needs of students with ADHD. This claim by Graczyk et al. (2005) is interesting when we recall that in research into teachers' knowledge of ADHD one of the weak points of the majority of the members of this group is their knowledge of the treatment of ADHD (Jarque et al., 2007; Ohan et al., 2008; Sciutto et al., 2000; Vereb & DiPerna, 2004; West et al., 2005). This lack of knowledge could lead to these teachers not having confidence in their abilities and so have more negative behavior and attitudes towards students with ADHD. As was pointed out by Bekle (2004), teachers must be aware that students with ADHD suffer from serious disadvantages in the classroom and they also have to have appropriate knowledge of the condition in order not to exert a negative influence on them. This same author holds that teachers with little or no knowledge of the disorder demand less from students suffering from it and approach them and praise them less, as well as criticize them more than other students.

Various studies have shown that teachers, in general, have a moderate level of knowledge of ADHD and that is necessary that this level of knowledge be increased (Ghanizadeh et al., 2006; Grazyk et al., 2005; Jarque et al., 2007; Kos, Richdale & Jackson, 2004; Sciutto et al., 2000; West et al., 2005; White et al., 2011). Results vary depending on the research consulted. For example, some studies have shown the average number of correct answers by teachers answering questionnaires about ADHD to be 80% (Barbaresi & Olsen, 1998; Bekle, 2004; Jerome, Gordon & Hustler, 1994; Jones & Chronis-Tuscano, 2008; Ohan et al., 2008), while other studies show the average percentage of correct answers as not exceeding 53% (Jarque et al., 2007; Kos et al., 2004; Sciutto et al., 2000; Stacey, 2003; West et al., 2005). These differences are probably due to the different methodologies used in the studies. The pioneering studies of teachers' knowledge of ADHD (Barbaresi & Olsen, 1998; Jerome et al., 1994) had an average of 24 questions to be answered on a True or False basis. Another series of authors (Bekle, 2004; Jones & Chronis-Tuscano, 2008; Ohan et al., 2008) used similar measurement instruments with the objective of their studies being as similar as possible to the earliest research on this matter. However, the results obtained from questionnaires with dichotomous response formats can be deceptive as they invite participants with doubts to opt for one or other possible answer and this contributes to an increase in the proportion of correct answers due to the divination effect (Jarque et al., 2007; Kos et al., 2006; Sciutto et al., 2000).

Sciutto et al. (2000) established the beginning of a new stage by creating a questionnaire with 36 items which included 3 dimensions and an answer format with three options: "True", "False" and "I don't know". It was also the first tool for measuring the knowledge of teachers of ADHD which had its indices of reliability and validity published. This tool has been a source of inspiration for more recent studies (for example, Guerra, 2010; Jarque et al., 2007; Stacey, 2003; West et al., 2005) which found more moderate levels of knowledge among teachers than those found in the earliest research. The increase in the number of questionnaire items with respect to the number used in the pioneering studies had an impact on the reduction of the number of right answers by respondents but it was the introduction of the three option response format which allowed for the obtaining of more exhaustive information, allowing the areas where the teachers had the most knowledge, the areas where they had the least and the areas where respondents made the most mistakes to be identified (Jarque & Tárraga., 2008; Kos et al., 2006; Sciutto et al., 2000). Another series of studies used either a Likert format with 5 options (Snider, Busch & Arrowood, 2003) or multiple choice questions (González-Acosta, 2006; Niznik, 2005; White, 2011) and the results obtained were similar to those found by three option format questionnaires.

To summarize, the differences found in the results of research into educators' knowledge of ADHD can be explained by the following factors: 1) at the world level there exist a fair number of studies of educators' level of knowledge of ADHD, nevertheless there is a shortage of instruments with which to accurately measure that knowledge; 2) the differences in the number of items making up the various questionnaires and their contents led to a variance in the number of right answers in the studies; 3) the response formats used were different; 4) the breadth of the sample varied considerably from one study to another, with some studies having a large number of participants while other had relatively few; and 5) although they were not taken into account in all the studies, such socio-demographic variables as the previous training received by the teachers with respect to ADHD or their previous experience in teaching children with this condition, had a significant influence on the results obtained. All of these factors must be taken into account when it comes to interpreting, comparing and generalizing the results obtained from these studies.

What emerges from these studies is that the teachers had limited training with regard to ADHD. It must be pointed out that many teachers suffer from either a lack of knowledge or false beliefs regarding the consequences, causes and treatment of ADHD (Pfiffner, 1999). Given this fact, Moreno (2008a) considers that in order to properly deal with students with ADHD, teachers should acquire information about the characteristics, implications and effective methods for dealing with this disorder. In the same way, Arcia et al. (2000) indicates that teachers need more rigorous training with regard to the causes of ADHD, the identification of its characteristics and the employment of strategies to deal with it that are effective in the classroom. Furthermore, arising from the weaknesses identified, in the bulk of the research the teachers involved explicitly expressed a desire to receive more information regarding ADHD (Bussing, Gary, Leon & Garvan, 2002; Jerome et al., 1994; Sciutto et al., 2000).

Today anyone who cares to look for it can find a wide variety of information about ADHD because of the large number of specialist publications and websites that deal with it. However, as Moreno (2008a) points out, it would be appropriate for teachers to develop their knowledge in a more formal and regulated manner such as by attending relevant lectures at congresses, symposiums, etc. and/or participating in courses, workshops and seminars that deal with the disorder. Furthermore, turning to work colleagues for advice and information can be a risky strategy as these colleagues might provide insufficient or false information. As Kos et al. (2006) have shown, those teachers who consider their level of knowledge of ADHD to be optimal do not search for additional information about it. Those who believe themselves to be ignorant about this topic, by contrast, do seek information about it. For this reason it is important for educators to be aware of their real level of knowledge regarding ADHD as well as the possible consequences of gaps in their knowledge and/or false beliefs about this disorder.

According to Sherman et al. (2008), the relevant variables that distinguish the teacher specially trained to assist and support students with ADHD are patience, knowledge of effective intervention techniques, the ability to collaborate with a multidisciplinary team, the use of gestures to communicate with students and a positive attitude towards children with special educational needs. In the same vein authors such as Pfiffner (1999) and Cooper and Bilton (2002) highlight two teacher factors that are key to the success of students with ADHD: knowledge and education about the disorder, and a positive attitude towards and perception of the recommended psychological and educational interventions. However, as noted by Kos et al. (2006), this line of research is relatively recent and, so far, it has produced few relevant studies in this regard. It is to be hoped that future studies will provide more clues about the characteristics that teachers ought to have in order to facilitate the academic success of children with ADHD.

Leaving aside the personal characteristics of teachers and focusing on the level of knowledge they possess on ADHD, several studies suggest that a significant percentage of teachers hold false beliefs and have gaps in their knowledge regarding ADHD and that, in addition, on many occasions they express negative attitudes towards children with this condition (for example, Barbaresi & Olsen, 1998; Jarque et al. 2007; Sciutto et al., 2000). As we have already indicated, change is possible through training. The problem is that there are few studies that examine programs designed to increase teachers' knowledge of ADHD and the bulk of those that exist have not analyzed the effectiveness of these programs or their methodological rigor. Barbaresi and Olsen (1998) pioneered the evaluation of training programs for teachers on ADHD. They developed a 27 item questionnaire with a dichotomous response format and administered it to a group of 44 teachers from the same school, both before they participated in a training program on ADHD and one month after its completion. The results of the study show that teachers' knowledge increased considerably after the intervention. However, it must be pointed out that Barbaresi and Olsen's (1998) study suffers from some methodological problems. In fact, the psychometric properties of the instrument used to measure knowledge of teachers were not adequate. Furthermore, the study was carried out in a single school and with a very small number of teachers.

The work of Jones et al. (2008) marked a significant change in the degree of scientific rigor of studies of the effectiveness of programs to educate teachers about ADHD. It controlled for several variables and used a random sample. The sample consisted of 142 primary school teachers from 6 different schools and had a control and an experimental group. The training given by Jones et al. (2008) to the teachers included a general overview of ADHD (identification and diagnosis) as well as information on the evidence based treatment of ADHD (pharmacological and psychosocial) and the theory and implementation of specific strategies for behavior management in the classroom. The level of teachers' knowledge of ADHD was evaluated before receiving the training and one month after it started. The authors of the study drew up a questionnaire for this purpose. The results indicate that teachers increased their knowledge of the disorder in some small measure but were very satisfied with the training. However, the instrument employed by Jones et al. (2008) to measure teachers' knowledge did not have adequate psychometric properties and the response format used in the questionnaire was dichotomous, all of which placed limits on the results of the study.

Syed and Hussein (2009) also carried out research on the effectiveness of a training program on ADHD for teachers. They carried out a pilot study in which they gave training to 49 teachers over the course of a week with 10 teaching contact hours. The training dealt with various topics related to ADHD. The teachers' knowledge was measured before the program began, just after it finished and six months later. The teachers' knowledge increased significantly and remained relatively stable over time. However, the instrument employed by the authors to measure the knowledge of the teachers, the questionnaire of Jerome et al. (1994), has not been validated and the response format used was dichotomous, all of which, once again, placed limits on the results of the study.

On the basis of what we have just discussed the necessity to continue to examine the effectiveness of programs aimed at improving teachers' knowledge of ADHD can be seen. The starting point for achieving this objective is that teachers be involved in training programs. In principle, any training received by the teacher will be beneficial for students with ADHD, as long as the teacher is able to transfer what he or she has learned to the classroom.

In short, teachers play a crucial role with respect to children with ADHD. As pointed out by Arcia et al. (2000), teachers need to identify children with possible ADHD, refer them to an appropriate specialist for evaluation, provide the specialist with valid and reliable reports about the skills, academic achievement and behavior of the children, and implement the necessary treatment strategies in the classroom. All this work could prove to be arduous if the teacher lacks previous knowledge of the matter. Thus, in order to examine the level of teachers' knowledge and offer appropriate training to address the problems detected, the following two lines of work need to be undertaken: the development and validation of instruments with appropriate psychometric properties able to accurately measure teachers' knowledge of ADHD, and the design, implementation and evaluation of training programs on ADHD for teachers.

### 1.2 The content validity of the questionnaires

When in social sciences or in any other discipline, a questionnaire is created with a specific goal, such as analyzing the level of infant and primary teachers' knowledge of ADHD, it is important to properly define exactly what is being evaluated. To obtain good results, that is to say, a good measurement of the construct, it is necessary to pay special attention to the first phase of the development of the instrument or, which amounts to the same things, to the content validity (for more on this concept such articles as those by McGartland, Berg-Weger, Tebb, Lee and Rauch, 2003 and McKenzie, Wood, Kotecki, Clark and Brey, 1999 can be consulted, or more technical pieces such as those by Haynes, Richard and Kubany, 1995 or Sireci, 1998).

Haynes et al. (1995) define content validity as the degree to which the elements of an assessment instrument are relevant and representative of the construct or concept to be evaluated. Following on from that description, content validity covers all elements of an instrument affecting the data collection, i.e. it includes aspects such as the content of the items, the form of presentation of the instrument, the instructions provided to participants, the estimated time for completion of the task, the codes of conduct during the application of the test, or the response format of the questionnaire.

The main steps to be taken in the creation of a new questionnaire in order to attain an appropriate degree of content validity are the following: 1) develop items based on the guidelines provided by expert researchers in the field; 2) achieve a format for the presentation of the questionnaire that is as neutral and agreeable as possible; 3) give clear and precise instructions to respondents, with no possibility of errors or dual interpretations; 4) develop tests whose estimated response times are short; 5) act in as neutral a manner as possible and without distracting the attention of the participants during the administration of the test; 6) select the answer format in accordance with the construct to be evaluated; 7) conduct an evaluation of the instrument by a panel of experts in the area and by a series of members of the population to whom the questionnaire is directed; and 8) make the relevant quantitative and qualitative analyses to enable the test to be refined and prepared for the next stage of validation.

Content validity is an indispensable prerequisite for the establishment of other types of validity. Any assessment instrument should pay special attention to content validity because it is a predictor of construct validity (Haynes et al., 1995) and serves as a preliminary analysis of factorial validity (McGartland et al., 2003).

As noted by Haynes et al. (1995), inferences derived from measurement instruments with unsatisfactory content validity must be doubted, even when other indicators of validity are satisfactory. In the case of questionnaires that have been developed to assess the level of knowledge of teachers regarding ADHD, little information regarding their content validity has been provided. There are few studies that mention this aspect of validity, and when it is mentioned, is made of it this occurs in the form of vague allusions to some of its component parts (drawing up of items and response format in particular). In general, most of the instruments have been constructed for the particular study itself and the strategy for their preparation has not been specified.

Throughout this chapter the process of drawing up the two language versions of a questionnaire created to measure the level of infant and primary teachers' knowledge of ADHD is described. This account has been prepared in accordance with the empirical evidence regarding ADHD and following the guidelines set for psychometric development and validation of questionnaires. In the case of the first study an initial version of the questionnaire was drawn up and contact then made with a group of experts in ADHD in order to provide the instrument with adequate content validity. In the case of the second study, the pilot test, a group of infant and primary school teachers, members of the group to which the questionnaire was aimed, completed it and evaluated it with the aim of carrying out a preliminary evaluation of the items which made it up.

## 2. Study 1: Drawing up the questionnaire

The aim of this first study was to establish the basis for the development of two language versions of the same questionnaire based on their content validity. This instrument was created to measure the level of knowledge of ADHD of teachers in infant and primary education. The questionnaire development process took into account the guidance of several experts regarding the content validity.

### 2.1 Method

#### 2.1.1 Participants

The group of 16 (N=8 for each language version) experts who participated in the study was formed by four doctors of psychology, all of them university lecturers from different fields of knowledge (Personality, Evaluation and Psychological Treatment, and Developmental Psychology and Education), a university professor of pediatrics, two child and youth psychiatrists working in hospitals, two outpatient care pediatricians, three psychologists and a education expert who are members of various associations of families of children with ADHD, an educational psychologist in private practice and two educational psychologists working in schools.

#### 2.1.2 Materials

A letter explaining the main aspects of the research and a questionnaire divided into 3 sections for the evaluation process by experts were drawn up. In the first section the experts were asked to evaluate the suitability of each of the 105 preliminary statements (75 positive and 30 negative) which had been drawn up to measure the knowledge of ADHD of infant and primary school teachers using a Likert scale with 4 response options (1= Unsuitable, 2= Barely suitable, 3= Fairly suitable, and 4= Very suitable). The second section requested proposals for changes to the questionnaire, if changes were thought to be necessary. In the third and final section the experts had to assign each item to one of 4 categories: 1) General information, 2) Symptoms/Diagnosis, 3) Etiology, and 4) Treatment. Finally in this section, the experts were asked if they wished to add a new category or eliminate any of the existing categories.

### 2.1.3 Procedure

First, the questionnaire items were developed. Based on the scientific literature it was decided that these statements should evaluate 4 categories: 1) General information (about the nature, prevalence, evolution and the problems associated with ADHD), 2) Symptoms/Diagnosis (the different symptoms that may occur in children with ADHD, as well as information on the process of diagnosing the disorder), 3) Etiology (cause or possible causes that influence the occurrence of ADHD), and 4) Treatment (the different types of interventions carried out with children with ADHD). On the basis of these categories and on guidelines provided by Barbero (2003), Moreno, Martínez and Muñiz (2004, 2006) and Hernández, Fernández-Collado and Baptista (2006) for the construction of items, 105 preliminary propositions were drawn up, 75 positive and 30 negative. The items were created from a comprehensive review of the literature on ADHD. Furthermore, in the light of the linguistic reality of the target population, the questionnaire was created in the two official languages of the Autonomous Community of the Basque Country: Spanish and Basque, so that each teacher could select either of the two languages to answer it.

Regarding the response format of the questionnaire, it was decided to use three-option response format. The criteria for selection can be found in the following section.

Once the items and response format had been decided on, the experts were contacted by telephone and asked about their availability for collaborating with the study. The experts who agreed to participate were sent an explanatory e-mail about the study along with the questionnaire, which has been described in the section on materials. They were given a period of two weeks to fill out the three sections of the questionnaire.

## 2.2 Results

An initial selection was made of the propositions considered to be "Very suitable" or "Fairly suitable" by at least 80% of the experts and "Unsuitable" by no more that 6% of them. This initial selection produced 64 items among which, at the suggestion of the experts, 7 were redrafted.

It should be noted that some experts suggested the elimination of the negative items or those that included information about the etiology and treatment of ADHD. These suggestions were not acted on. In fact, based on the literature on questionnaires that measure teachers' knowledge of ADHD, it seems that the presence of true and false statements may be desirable to avoid response tendencies or repetitions, which would lessen the validity and reliability of scores (Sciutto et al., 2000; West et al., 2005). Expressed in another way, the use of both negative and positive propositions is recommended to avoid both acquiescence bias (the tendency to agree with what is stated in an item, regardless of its content) and affirmation bias (the tendency to give the answer the participant thinks the researcher wants to receive) (Martínez Arias, Hernández Lloreda & Hernández Lloreda, 2006). In addition, based on the evidence that teachers play a key role in identifying and treating children with ADHD (Jarque et al. 2007; Ohan et al., 2008; Snider et al. 2003; Vereb & DiPerna, 2004) it was considered important that this group have sufficient knowledge of the etiology and treatment of this disorder. For these reasons and in spite of the contrary opinion of some experts, it was decided to include 12 items in various categories, principally "General information" and "Etiology" of which 3 were true and 9 false.

As a result of the selection process outlined above, a questionnaire with 76 propositions was obtained. The fundamental criteria employed for assigning each of the propositions to one of the four dimensions was the following: that at least 70% of the experts had placed that item in that category. Thus the distribution of items by category turned out as follows: the "General information" category received 23 items, the "Symptoms/Diagnosis" category received 30 items, the "Etiology" category received 7 items and, finally, the "Treatment" category received 16 items. Of the 76 propositions 59 were positive and 17 negative.

With regard to the response format, it has already been mentioned that most studies that have analyzed the level of knowledge of teachers regarding ADHD have used questionnaires with two possible answers (True, False) or with three response options (True, False, I don't know). The pioneering studies of teachers' knowledge of ADHD opted for the two response options format, True or False. However, the scores obtained using this format are usually distorted because participants have a high probability of getting the correct answer by chance, that is, participants are obliged to have to select one of the two options provided, which increases the level of correct answers obtained by guessing (Jarque et al. 2007; Martínez Arias et al. 2006; Sciutto et al., 2000). The fundamental problem that arises from this is that it produces an increase in the error variance of the scores and so reduces the reliability and validity of the questionnaire (Martínez Arias et al., 2006).

This limitation involved in dichotomous response format was dealt with by the introduction of the three-option response format: True, False, I don't know. The existence of a third option allows for more reliable information to be obtained about teachers' knowledge of ADHD (Jarque et al., 2007; Kos et al., 2006; Sciutto et al., 2000), given that in the case of doubt, the participant is not obliged to choose between a negative and positive response.

As was pointed out by Muñiz (2001), the quality of questionnaires improves as the number of response alternatives is increased, because the increase reduces the probability of getting the right answer by chance. Thus, the three-format options, by comparison with the two option ones, increase the probability that participants' responses will accurately reflect their knowledge and not the divination effect, and also allow participants to express their position more precisely, as well as providing a higher degree of reliability. For these reasons and for those previously set out, in this study it was decided to use a three-option response format (True, False, I don't know). A response format with an uneven number of options was chosen, that is, one with an intermediate point, because it was considered necessary that the format reflect the gaps in teachers' knowledge of ADHD, as well as their knowledge of and false ideas regarding it.

## 3. Study 2: Pilot study

This second study was designed to improve the questionnaire created in the first study conducting a pilot application of the instrument to the target population. The aim was to obtain preliminary information about the functioning of items and a revision of the formal aspects of the questionnaire.

### 3.1 Method

#### 3.1.1 Participants

166 teachers participated in this study (N=68 for the Spanish version and N=98 for the Basque version), 136 women and 30 men. These teachers worked at 17 infant and primary schools in the Autonomous Community of the Basque Country and Navarre. The average age of participants was 41, though the ages of teachers covered a wide range between 21 and 61. The teachers had an average of 16 years (*SD*=11) in the profession (3 teachers did not provide this information) and had an average of 53 students each in the current school year (*SD*=80), although it should be noted that in this respect there was great variability because some of the teachers taught several groups of students. Most of the teachers were working in one specialized area but a few were specialists in various areas. Thus, 50% worked as teachers in primary education, 30% in infant education, 10% were special educational needs teachers, 8% were foreign language teachers, 6% were music education teachers and the remaining 6% were physical education teachers.

49% stated that they had not received any specific training regarding ADHD, while 51% had received some. Among those who had received training, the average duration of such training was 14.6 hours (*SD*=19.35), however, there was great variability in the total duration of the training received by teachers. The data collected indicated that most of the teachers who had received specific training on ADHD had received it at conferences (29%), followed by a course forming part of a teacher training studies (19%), continuing education courses (15%), from associations of families of children with ADHD (12%) and postgraduate or master's degree courses (4%). In addition, 85% of the teachers confirmed having received some information about ADHD from various sources. These sources of information included the media (46%), people around the teacher (family, friends, coworkers, etc.) (40%), professionals from outside the schools (31%), books (27%), scientific or professional journals (20%) and associations of families of children with ADHD (17%). It should be pointed out that 62 people did not specify the exact source through which they had received information about ADHD.

58% of the teachers recognized having had a child diagnosed with ADHD in their classroom at some point, adding up to an average of approximately 2 children (*SD*=5.8) per teacher over the course of their professional careers. Moreover, 40% of the participants claimed to know someone with ADHD outside their workplace. Furthermore, on a scale of 1 to 10 points, the teachers placed their level of knowledge of, and teaching ability with regard to ADHD at an average of 4 and 4.26 respectively (*SD*=1.73 and *SD*=1.86). On this scale 1 corresponded to the lower end, which means that the teacher had no knowledge regarding ADHD and did not feel prepared to appropriately teach a child with ADHD, while 10 corresponded to the upper end, which implies that the teacher had an excellent knowledge regarding ADHD and felt fully prepared to appropriately teach a child with ADHD.

### 3.1.2 Materials

Teachers who participated in this study completed a questionnaire divided into three sections. In the first section, referring to socio-demographic data, teachers had to provide certain information about themselves: age, gender, professional specialty, specific training received on ADHD or personal experience with individuals with ADHD, etc. The second section aimed to assess the knowledge level of teachers in infant and elementary schools with respect to ADHD and had 76 items, 59 of which were positive and 17 negative, with a three-option response format (T= True, F= False, DK= I don't know). The third and final section of the questionnaire asked participants their personal opinion on the following issues: Their understanding of the instructions, the wording of the items, the response format, the length and duration of the questionnaire and their general opinion of the test they had just completed. The three sections of the questionnaire were identical for the two language versions. The average time stipulated for completing the questionnaire was 20 minutes.

### 3.1.3 Procedure

The managements of the 38 schools that were selected to participate in research were contacted by telephone. Each center was provided with general information about the project and offered the opportunity to participate in it. Of the 38 schools selected, 17 chose to involve themselves in the study. One of the members of the research group visited each center and explained the procedure for the application of the questionnaire. The member of the research group also provided the school management with a dossier containing all the information regarding the project, an informative letter addressed to each teacher with an informed consent declaration attached and two copies of the questionnaire, one in Spanish and one in Basque, for each teacher, so that it would be the teacher him or herself who would choose in which language to respond. Also, all the teachers were informed of the voluntary nature of their participation in the project and the confidential nature of their responses.

Each school had a period of between a week and 15 days for the application and collection of the test from the teachers. At the end of this period a member of the team went back to each school to collect the completed tests.

The schools were also informed that once the data were analyzed the research team would again contact the schools that had participated in the study to inform them of the global results of the study.

## 3.2 Results

In the first place, quantitative analysis of the items was carried out. In order to obtain a group of statements that maximized the variance of the questionnaire and to increase the internal consistency of each category, we selected those clauses with high discriminatory power (with values equal to or greater than 0.30 in the corrected correlation coefficient between the score on each of the 76 items and the total score in the category to which they belonged), and a considerable standard deviation. As a result of this analysis 25 of the statements were removed.

After eliminating items that had a low correlation with other items of the same category, it can be seen that the four categories proposed in the questionnaire had generally acceptable Cronbach's alpha coefficients for the two language versions of the test, with coefficients ranging between 0.55 and 0.85 for the Basque version and between 0.67 and 0.84 for the Spanish version. Table 1 shows the descriptive statistics and Cronbach's alpha coefficients of the categories and of the entire questionnaire for each language version. It should be noted that with SPSS 18.0 software, the correct answers were coded with 1 point and the wrong ones and the gaps in knowledge with 0.

As can be seen in Table 1, the alpha coefficients are generally better for the Spanish version. This difference is especially noticeable in the "Etiology" category, a dimension that has a moderate Cronbach's alpha coefficient for the version in Basque. We believe that this difference in the alpha coefficient values between the different language versions may be due to an error in the wording of the items in the Basque version, since both versions were, in theory identical.

| Questionnaire Categories | Basque Language Version N=98 | | Spanish Language Version N=68 | |
|---|---|---|---|---|
| General Information (13 items) | M | 7.64 | M | 8.63 |
| | SD | 3.12 | SD | 3.27 |
| | Alpha | 0.77 | Alpha | 0.81 |
| Symptoms/Diagnosis (21 items) | M | 15.28 | M | 16.90 |
| | SD | 4.17 | SD | 3.81 |
| | Alpha | 0.85 | Alpha | 0.84 |
| Etiology (4 items) | M | 1.40 | M | 2.16 |
| | SD | 1.20 | SD | 1.52 |
| | Alpha | 0.55 | Alpha | 0.76 |
| Treatment (13 items) | M | 9.55 | M | 10.63 |
| | SD | 2.61 | SD | 2.07 |
| | Alpha | 0.76 | Alpha | 0.67 |
| TOTAL (51 items) | M | 33.88 | M | 38.32 |
| | SD | 8.93 | SD | 8.76 |
| | Alpha | 0.90 | Alpha | 0.91 |

Table 1. Descriptive statistics and Cronbach's alpha for the two language versions of the questionnaire.

In order to know different opinions about the design and clarity of the questionnaire items, we proceeded to the qualitative analysis of the contributions of members of the target population. After examining the views of teachers, it was decided to change the wording of 8 statements, 6 in the version in Basque and 2 in the Spanish version. Respondents to the version in Basque indicated that there was an ambiguous term repeated in four of the items "Etiology" category, something which had not happened in the Spanish version. The word that caused the problem was "eragin", which in Basque can mean both "cause" and "influence". Therefore, it was decided to reformulate those four statements in the Basque version of the questionnaire. Moreover, it was also decided to change the wording of two sentences in both versions of the questionnaire because a significant percentage of teachers who participated in the study reported having difficulties understanding two items in the categories "Symptoms/Diagnosis" and "Treatment". More specifically, a significant percentage of the teachers reported that they were unable to understand the following statement, "Children with ADHD have difficulties in delaying the receiving of rewards", particularly the phrase "delaying the receiving of rewards". The teachers also had difficulty understanding this statement, "Behavior modification techniques are one of the most common treatments for children with ADHD", particularly the phrase "behavior modification". These two problems were considered and the two items, from the "Symptoms/Diagnosis" and "Treatment" categories, were revised and reformulated for both language versions of the questionnaire.

The final distribution of items by category was as follows: the "General information" category had 13 items, the "Symptoms/Diagnosis" category had 21 items, the "Etiology" category had 4 items and the "Treatment" category had 13 items. The definitive version of the questionnaire had 51 randomly distributed items with not more than two consecutive items from the same category.

In the final version of the questionnaire 45 of the items were positive and 6 were negative. Thus, the possible scores ranged from 0, for the minimum level of knowledge, to 51, for the maximum level.

Table 2 presents the versions of the questionnaire in Basque and Spanish as well as their English translation. All the items are located in their corresponding categories.

| **General Information (13 items)** | |
|---|---|
| B | AGHNa duten mutil eta nesken proportzioa antzerakoa da. |
| S | La proporción de varones y mujeres con TDAH es similar. |
| E | The proportion of men and women with ADHD is similar. |
| B | AGHNa duten gazteek, gainerako gazteek baino maizago utzi ohi dituzte ikasketak. |
| S | Los/as jóvenes con TDAH abandonan los estudios con una mayor frecuencia que el resto de jóvenes. |
| E | Young people with ADHD drop out of school more frequently than others. |
| B | AGHNa oso estresagarria da haurrarekin bizi diren pertsonentzat. |
| S | El TDAH resulta muy estresante para las personas que conviven con el/la niño/a. |
| E | ADHD is stressful for the people living with the child suffering from it. |
| B | Normalean, familiartekoek, irakasleek eta ikaskideek AGHNa duen haurraren jokabide negatiboei eskaintzen diete arreta eta gutxitan positiboei. |
| S | Normalmente, los familiares, maestros/as y compañeros/as suelen prestar atención a las conductas negativas del/a niño/a con TDAH pero raras veces a las positivas. |
| E | Typically, family members, teachers and classmates tend to pay attention to negative behaviors of the child with ADHD but rarely to positive behaviors. |
| B | AGHNa duten haurrek autoritate irudiekiko (gurasoak, irakasleak, e.a.) jokabide bihurria eta etsaitasuna aurkeztu dezakete. |
| S | Los/as niños/as con TDAH pueden presentar un comportamiento rebelde y hostil hacia las figuras de autoridad (progenitores, maestros/as, etc.). |
| E | Children with ADHD may behave in a rebellious and hostile manner towards authority figures (parents, teachers, etc.). |

| | |
|---|---|
| B | AGHNa duen haur bat izateak, familia baten bizitza soziala mugatu dezake. |
| S | Tener un hijo o una hija con TDAH puede limitar la vida social de una familia. |
| E | Having a son or a daughter with ADHD can limit the social life of a family. |
| B | AGHNa duten haurrek ez dute azterketetan benetan dakitena erakusten. |
| S | Los/as niños/as con TDAH no suelen reflejar en los exámenes lo que realmente saben. |
| E | The examination results of children with ADHD usually do not reflect their real level of knowledge. |
| B | AGHNa duten mutilek, neskek baino hiperaktibitate-inpultsibitate maila handiagoa aurkeztu ohi dute. |
| S | Los varones con TDAH suelen presentar un mayor grado de hiperactividad-impulsividad que las mujeres. |
| E | Boys with ADHD tend to have a greater degree of hyperactivity-impulsivity than girls. |
| B | AGHNa duten haurrek, irakurketa, idazketa eta kalkuluarekin zailtasunak izaten dituzte. |
| S | Los/as niños/as con TDAH suelen tener dificultades con la lectura, la escritura y el cálculo. |
| E | Children with ADHD often have difficulty with reading, writing and arithmetic. |
| B | AGHNa duten haurrek, gainerako haurrek baino istripu, erortze eta lesio gehiago izaten dituzte. |
| S | Los/as niños/as con TDAH tienden a sufrir más accidentes, caídas y lesiones que el resto de niños/as. |
| E | Children with ADHD are more prone to accidents, falls and injuries than other children. |
| B | AGHNa duten haurrek euren guraso eta irakasleengandik maiz jasotzen dituzte kritikak, mehatxuak eta zigorrak. |
| S | Los/as niños/as con TDAH suelen recibir críticas, amenazas y castigos frecuentes por parte de sus progenitores y maestros/as. |
| E | Children with ADHD often receive criticism, threats and punishments from their parents and teachers. |
| B | AGHNa duten haurren kopuru garrantzitsu batek kideekin harreman txarrak izaten dituzte. |
| S | Un número importante de niños/as con TDAH suele tener malas relaciones con los/as compañeros/as. |
| E | A significant number of children with ADHD tend to have bad relations with their classmates. |
| B | Oro har, nerabezaroan gehiegizko aktibitatea gutxitu egiten da, baina inpultsibitatea eta arreta-zailtasunak mantendu egiten dira. |
| S | En general, en la adolescencia disminuye el exceso de actividad, aunque la impulsividad y las dificultades atencionales se mantienen. |
| E | In general, excessive activity reduces in adolescence, though the impulsiveness and difficulties with attention remain. |

| Symptoms/Diagnosis (21 items) | |
|---|---|
| B | AGHNa duten haurrek besteen ekintzak eteten dituzte edota horietan sartzen dira. |
| S | Los/as niños/as con TDAH interrumpen o se inmiscuyen en las actividades de otros/as. |
| E | Children with ADHD interrupt or intrude on the activities of others. |
| B | AGHNa duten haurrek gauzak maiz ahazten dituzte. |
| S | Los/as niños/as con TDAH suelen olvidarse de las cosas frecuentemente. |
| E | Children with ADHD often tend to forget things. |
| B | AGHNaren oinarrizko sintomak dira gehiegizko aktibitatea, arreta-zailtasunak eta inpultsibitatea. |
| S | Los síntomas básicos del TDAH son el exceso de actividad, los problemas de atención y la impulsividad. |
| E | The core symptoms of ADHD are excessive activity, attention problems and impulsiveness. |
| B | AGHNa duten haurrek badirudi ez dutela entzuten hitz egiten zaienean. |
| S | Los/as niños/as con TDAH parece que no escuchan cuando se les habla. |
| E | Children with ADHD seem not to listen when spoken to. |
| B | AGHNa duten haurrek, egiten dituzten ekintzengatik berehala sarituak izatea dute gustuko, horrela izan ezean erraz desmotibatu daitezke. |
| S | A los/as niños/as con TDAH les gusta ser recompensados/as de forma inmediata por sus acciones, de lo contrario pueden desmotivarse con facilidad. |
| E | Children with ADHD like to be immediately rewarded for their actions and any delay in receiving the reward tends to demotivate them. |
| B | AGHNa duten haurrek euren ekintzen ondorioetan pentsatu gabe jarduten dute. |
| S | Los/as niños/as con TDAH actúan sin pensar en las consecuencias de sus acciones. |
| E | Children with ADHD act without thinking through the consequences of their actions. |
| B | AGHNa duten haurrek gelako arauak eta arau sozialak errespetatzeko zailtasunak izaten dituzte. |
| S | Los/as niños/as con TDAH presentan dificultades para respetar las normas del aula y las normas sociales. |
| E | Children with ADHD have difficulty complying with the rules of the classroom and social norms. |
| B | AGHNa duten haurrek denboraren antolaketa eta banaketarekin zailtasunak izaten dituzte: lehentasunak finkatzea kosta egiten zaie, zereginak amaitzeko denbora falta izaten dute, hitzorduetara berandu iristen dira, e.a. |
| S | Los/as niños/as con TDAH suelen tener dificultades de organización y distribución del tiempo: les cuesta establecer prioridades, les falta tiempo para acabar las tareas, llegan tarde a las citas, etc. |
| E | Children with ADHD usually have problems with organization and distribution of time: They have difficulties in establishing priorities, they don't have time to finish their tasks, arrive late for appointments, etc. |
| B | AGHNaren diagnostikoa prozesu konplexua da eta, ahal bada, medikuntza, psikologia eta hezkuntza bezalako ezagutza arlo desberdinetako profesionalek osatutako ekipo batek egin beharrekoa. |

| | |
|---|---|
| S | El diagnóstico del TDAH es un proceso complejo que debe ser realizado, preferiblemente, por un equipo de profesionales de distintos ámbitos como la medicina, la psicología y la educación. |
| E | The diagnosis of ADHD is a complex process that must be performed by a team of professionals from various fields such as medicine, psychology and education. |
| B | AGHNa duten haurrei kosta egiten zaie hasi dituzten lanak amaitzea. |
| S | A los/as niños/as con TDAH les cuesta finalizar las tareas que han iniciado. |
| E | Children with ADHD have trouble completing the tasks they have started. |
| B | AGHNa duten haurrek euren txanda itxaroteko zailtasunak dituzte. |
| S | Los/as niños/as con TDAH tienen dificultades para guardar su turno. |
| E | Children with ADHD have difficulty awaiting their turn. |
| B | AGHNa duten haurrek gehiegizko aktibitatea eta mugimendua aurkezten dute lasaitasuna eskatzen duten egoeratan: eserlekutik altxatzen dira, esku eta hankak mugitzen dituzte, korrika egiten dute, salto egiten dute, e.a. |
| S | Los/as niños/as con TDAH presentan un exceso de actividad y movimiento en situaciones que requieren calma: se levantan de su asiento, mueven manos y pies, corren, saltan, etc. |
| E | Children with ADHD suffer from an excess of activity and movement in situations that require calm: they get up from their seats, move their hands and feet, run, jump, etc. |
| B | AGHNa duten haurrek, arropa, eskola-materiala, jostailuak, eta antzerako beste objektuak maiz galtzen dituzte. |
| S | Los/as niños/as con TDAH pierden frecuentemente objetos como ropa, material escolar, juguetes, etc. |
| E | Children with ADHD often lose items such as clothing, school supplies, toys, etc. |
| B | Arreta-gabezia sintomak bakarrik aurkezten badira, zailagoa da AGHNa detektatzea. |
| S | Es más difícil detectar el TDAH si se presentan exclusivamente los síntomas de falta de atención. |
| E | It is more difficult to diagnose ADHD if the only symptoms presented relate to lack of attention. |
| B | AGHNa duten zenbait haurrek gehiegi hitz egiten dute. |
| S | Algunos/as niños/as con TDAH hablan en exceso. |
| E | Some children with ADHD talk too much. |
| B | AGHNa duten haurrak ez dira gaizki portatzen apropos, gertatzen dena da kosta egiten zaiela beraien jokabidea kontrolatzea. |
| S | Los niños/as con TDAH no se portan mal deliberadamente, lo que sucede es que les cuesta controlar su conducta. |
| E | Children with ADHD do not misbehave deliberately, what happens is they have difficulty controlling their behavior. |
| B | AGHNa duten haurrek euren emozioak kontrolatzeko zailtasunak izaten dituzte, batez ere haserrea. |
| S | Los/as niños/as con TDAH suelen tener dificultades para controlar sus emociones, sobre todo la rabia. |

| | |
|---|---|
| E | Children with ADHD often have difficulty controlling their emotions, especially anger. |
| B | AGHNa duten haurrek galdera amaitu baino lehen erantzuten dute. |
| S | Los/as niños/as con TDAH responden antes de que se les haya terminado de formular la pregunta. |
| E | Children with ADHD start to answer before their interlocutor finishes asking the question. |
| B | AGHNa duten haurrak buru-esfortzu jarraitua eskatzen duten lanak egin behar dituztenean kexatu egiten dira edota horiek egitea saihesten dute. |
| S | Los/as niños/as con TDAH se quejan cuando tienen que realizar tareas que requieren un esfuerzo mental continuo o evitan realizarlas. |
| E | Children with ADHD complain when they have to perform tasks that require sustained mental effort or try to avoid doing them. |
| B | AGHNa duten haurrek barne-mintzairaren garapenean atzerapenak izaten dituzte, beraien burutik pasatzen den guztia hitzez adierazteko joera dute eta eskolako lanak egiterakoan ozenki hitz egiteko joera dute. |
| S | Los/as niños/as con TDAH muestran un retraso en el desarrollo del lenguaje interno, acostumbran a verbalizar todo lo que pasa por su cabeza y suelen hablar cuando hacen los trabajos escolares. |
| E | Children with ADHD display a delayed development of inner speech, tend to verbalize everything that goes through their minds and talk while doing their schoolwork. |
| B | Irakasleek funtzio oso garrantzitsua betetzen dute AGHNaren detekzioan. |
| S | Los/as maestros/as juegan un papel muy importante en la detección del TDAH. |
| E | Teachers have a very important role in the detection of ADHD. |
| **Etiology (4 items)** | |
| B | Egungo bizitza erritmo estresagarriaren ondorioz haurrek AGHNa izan dezakete. |
| S | En los/as niños/as, el TDAH puede deberse al estrés generado por el actual ritmo de vida. |
| E | In children, ADHD may be caused by the stress generated by the current pace of life. |
| B | Gurasoen heziketa estilo autoritario eta kritikoaren ondorioz haurrek AGHNa izan dezakete. |
| S | Un estilo educativo autoritario y crítico por parte de los progenitores puede originar el TDAH. |
| E | A critical and authoritarian style of education by parents can lead to ADHD. |
| B | Gurasoen dibortzioa edo anai-arreba baten jaiotza bezalako gertaera estresagarrien ondorioz haurrek AGHNa garatu dezakete. |
| S | Sucesos estresantes, tales como el divorcio de los progenitores o el nacimiento de un/a nuevo/a hermano/a, pueden ser la causa del TDAH. |
| E | Stressful events such as divorce of parents or the birth of a new brother or sister may be the cause of ADHD. |
| B | Hezkuntza txarra edo familia-giro kaotikoaren ondorioz haurrek AGHNa izan dezakete. |

| | |
|---|---|
| S | Una mala educación o un ambiente familiar caótico pueden ser la causa del TDAH. |
| E | ADHD can be caused by a bad education or a chaotic home environment. |
| **Treatment (13 items)** | |
| B | AGHNaren tratamenduak urteak irauten ditu. |
| S | El tratamiento del TDAH se prolonga durante años. |
| E | ADHD treatment continues over a period of years. |
| B | AGHNa duen haurraren sintomak urteak pasa ahala eta inolako tratamendurik gabe hobera egingo dutela uste izateak, etorkizunean izan ditzakeen aukerak murrizten dizkio. |
| S | Esperar a que los síntomas del/a niño/a con TDAH mejoren con el paso de los años sin aplicarle ningún tratamiento, supone restarle oportunidades de futuro. |
| E | Waiting for the symptoms of ADHD to improve over the years, without any kind of treatment implies, means taking away opportunities for the child's future. |
| B | AGHNa duten haurrek espezialitate desberdinetako profesionalen laguntza behar izaten dute. |
| S | Los/as niños/as con TDAH necesitan ser atendidos/as por profesionales de distintas especialidades. |
| E | Children with ADHD need to be cared for by professionals from various specialties. |
| B | AGHNa duen haurraren guraso eta irakasleek teknika psikologikoen aplikazioan aktiboki parte-hartu behar dute. |
| S | Los progenitores y los/as maestros/as del/a niño/a con TDAH han de participar activamente en la aplicación de las técnicas psicológicas. |
| E | Parents and teachers of children with ADHD need to actively participate in the application of psychological techniques. |
| B | AGHNa duen ikaslea, irakaslea erraztasunez iritsi daitekeen ikasmahai batean kokatzea komeni da. |
| S | Conviene situar al alumno/a con TDAH en un pupitre de fácil acceso para el/la maestro/a. |
| E | It is appropriate to locate children with ADHD in desks to which the teacher has easy access. |
| B | Jokabidea aldatzeko teknikak (errefortzu positiboa, puntu sistema, denbora kanpora, e.a.) AGHNa duten haurrekin gehien erabiltzen diren esku-hartze tekniketako bat dira. |
| S | Las técnicas de modificación de conducta (refuerzo positivo, sistema de puntos, tiempo fuera, etc.) constituyen una de las intervenciones más utilizadas en niños/as con TDAH. |
| E | Behavior modification techniques (positive reinforcement, point system, timeout, etc.) constitute one of the most commonly used treatments for children with ADHD. |
| B | Ikasle-talde handian edota ikasle-talde txikian lan egin arren, AGHNa duten haurren eskola-errendimendua antzerakoa izaten da. |
| S | El rendimiento académico de los/as niños/as con TDAH suele ser similar cuando trabajan en grupos grandes de alumnos/as que cuando trabajan en grupos reducidos. |

| | |
|---|---|
| E | The academic performance of children with ADHD is usually similar when they work in big groups of students to what it is when they work in small groups. |
| B | AGHNari buruzko informazio orokorra jasotzeak, irakasle, familiarteko eta ikaskideek haur horiekiko dituzten jarrerak hobetzea dakar. |
| S | Recibir información general sobre el TDAH mejora la actitud de los/as profesores/as, familiares y compañeros/as hacia estos/as niños/as. |
| E | Receiving general information about ADHD improves the attitudes of teachers, parents and classmates towards children who suffer from it. |
| B | AGHNa duen haurraren eskola-errendimendua eta eskola-egokitzapena hobetu daitezke baldin eta bere irakasleek nahaste horri buruzko heziketa zehatza badute. |
| S | El rendimiento y la adaptación escolar del/a niño/a con TDAH pueden mejorar si sus maestros/as tienen formación específica en este trastorno. |
| E | The performance and school adjustment of children with ADHD may improve if teachers have specific training in this disorder. |
| B | AGHNa duten haurren arreta-gaitasuna hobetzeko teknika eta programa zehatzak existitzen dira. |
| S | Existen técnicas y programas específicos para mejorar la capacidad atencional de los/as niños/as con TDAH. |
| E | There are specific techniques and programs to improve the attention span of children with ADHD. |
| B | Irakasleak AGHNa duen ikaslea jasotzen ari den tratamenduan parte-hartzeak, terapiaren emaitzetan eragina du. |
| S | La colaboración del/a maestro/a en el tratamiento que está recibiendo el/la alumno/a con TDAH influye en los resultados de la terapia. |
| E | The collaboration of the teacher with the treatment being received by the child with ADHD has an influence on the result of that treatment. |
| B | Irakasleak AGHNa duen haurrari aukerak eskaini behar dizkio gehiegizko mugimendua bideratu dezan. |
| S | El/la maestro/a ha de ofrecer oportunidades para que el/la alumno/a con TDAH pueda canalizar el exceso de movimiento. |
| E | The teacher must provide opportunities for the student with ADHD to channel their excessive movement. |
| B | Autoaginduen teknikek (haurrak bere jokabidea bideratzeko bere buruari ematen dizkion jarraibide eta aginduak) AGHNa duten haurren inpultsibitatea gutxitzea ahalbidetzen dute. |
| S | Las técnicas de autoinstrucciones (instrucciones u órdenes que el/la niño/a se da a sí mismo/a para dirigir su conducta) permiten reducir la impulsividad de los/as niños/as con TDAH. |
| E | Self-instruction techniques (instructions that the child gives to him or herself to direct his or her behavior) can reduce the impulsiveness of children with ADHD. |

B: Basque; S: Spanish; E: English.

Table 2. Distribution of the questionnaire items by category in Basque and Spanish, with English translation.

In summary, the quantitative and qualitative analyses that were conducted suggested the modification of several components of the content validity proposed by Haynes et al. (1995). Changes were made in the drafting of eight items (two in the Spanish version and six in the Basque version) and, by eliminating 25 of the preliminary questionnaire items, the test duration was reduced significantly. Furthermore, it was decided to highlight the relevant information to which the participants had to respond, with the objective of reducing the number of missing values. However, the questionnaire instructions and the response format did not change because they were valued very positively by most participants.

## 4. Conclusions

Content validity is an essential element in the preparation of any research instrument. Authors such as McGartland et al. (2003) have stated that content validity is subjective, but fortunately it can be provided with objectivity by following a rigorous two-phase process (Beck & Gable, 2001; Lynn, 1986). The first phase deals with those issues that need to be taken into account to propose a first version of the instrument. In other words, it requires a precise conceptualization of the construct to be studied, the identification of its dimensions or categories, the generation of a sufficiently large and representative battery of items for each of the dimensions or categories, care in the presentation of the items, the writing of simple and precise instructions, a decision on the response format and an estimate of the time necessary to complete the test. The second phase consists of employing a panel of experts to obtain appropriate information in order to review the research instrument. The proper selection of experts is essential in this regard, as is care in the wording of the instructions given to them. Although some researchers downplay this two-step process, it often occurs that in cases where it is not used, the instrument turns out to be invalid once its psychometric properties have been analyzed. In order for this not to occur, content validity of needs to be given high priority in the process of developing a research instrument (Beck & Gable, 2001).

Usually, a well-crafted plan to ensure content validity leads to the obtaining of reliable results. However, despite the fact that content validity is one of the most critical steps in the preparation of any research instrument, it does not usually receive the attention it deserves. Polit and Beck (2006) assert that, in general, in studies on the development of research instruments, there is a need for greater transparency with regard to content validity. Something similar happens with the instruments that have been developed to assess the level of teachers' knowledge of ADHD. In recent decades many instruments have been developed for this purpose, however, a large portion of them have not been validated, and those instruments with acceptable psychometric properties show evidence of limited content validity.

In this context, the main objective of the present research was to develop a preliminary questionnaire with appropriate content validity. To do this, first, we took into account the definition of content validity proposed by Haynes et al. (1995) and a study was carried out to provide objectivity for the content validity process. The final result of first study was satisfactory, due in part to the participation of a considerable number of qualified experts. Secondly, a pilot study was conducted that enabled the research instrument to be refined. Through the carrying out of the pilot study it was possible to assess the functioning of items and discard or reformulate some of them. In addition, this preliminary study was useful for making adjustments in the socio-demographic data section of the questionnaire, and helped to identify difficulties that might arise during the administration of the questionnaire to a large and representative sample of the target population.

The authors of this chapter consider the work they have carried to be new in terms of its area of application. The level of teachers' knowledge of ADHD is a relatively recent field of research and most of the instruments created to assess this level of knowledge have shown evidence of limited content validity. Due to the importance of content validity, it is necessary place more focus on it, something we tried to do in this phase of the construction of our questionnaire.

As noted by Pfiffner (1999), many teachers lack knowledge or have false beliefs about the nature, course, consequences, causes and treatment of ADHD. Currently, KADDS (Knowledge of Attention Deficit Hyperactivity Disorder) (Sciutto et al., 2000) is one of the best and most widely used tool for the analysis of teachers' knowledge regarding ADHD. However, this questionnaire only measures three areas of knowledge related to ADHD: 1) Symptoms/Diagnosis of ADHD, 2) General information about the nature, causes and outcome of ADHD, and 3) Treatment of ADHD. The instrument that we have presented in this chapter covers a broader spectrum of teachers' knowledge of ADHD, as it proposes four categories for evaluation: 1) General information about ADHD, 2) Symptoms/Diagnosis of ADHD, 3) Etiology of ADHD, and, 4) Treatment of ADHD. With this instrument it is expected to be possible to obtain more detailed information about teachers' knowledge of ADHD.

Given that teachers play an important role in the academic success, and personal and social development of children with ADHD, and their level of knowledge regarding the disorder affects their behaviors and attitudes toward children who have this condition, we consider it necessary to have an assessment instrument that measures this level of knowledge in a rigorous way. We also believe that this instrument may be useful in detecting which aspects of the universe of ADHD teachers lack knowledge about or with regard to which they have false beliefs. The information so produced could form the basis for designing training materials and programs aimed at this group. In this regard it should be noted that in many cases the training offered to teachers is based on the intuition of the trainers. Consequently, the development and subsequent use of an instrument with adequate psychometric properties to assess the level of teachers' knowledge of ADHD can provide objective data that will permit the provision a response tailored to the actual training needs of teachers.

Furthermore, we believe that the results obtained from the instrument we are developing could also have a positive effect on the upgrading of undergraduate university studies in infant and primary education. Rigorous scientific contributions about one of the most common disorders among children may well be incorporated into the training curricula for future teachers.

Finally, it should be noted that instrument presented throughout this chapter is currently under a validation process. The validation of the instrument being developed requires the obtaining of different kinds of evidence of validity with a broad sample of participants. In order to do this, during its application phase the questionnaire will be administered to 1000 infant and primary school teachers in approximately 100 schools and teaching centers in the Basque Country. Factorial analysis will be carried out to determine its dimensionality. Furthermore, evidence of validity based on the relationship between the scores obtained in the questionnaire and socio-demographic variables such as specialities of the teachers, their general or specific training with regard to ADHD and their previous experience with students with ADHD, will be provided. The convergent validity of the instrument will be analyzed by examining the correlation between its scores with those obtained from the Spanish version of Knowledge of Attention Deficit Hyperactivity Disorder (KADDS) (Sciutto et al., 2000; Spanish version by Jarque et al., 2007). The reliability of the instrument will also be examined by an analysis of its internal consistency and temporal stability. All of this will serve to improve the instrument by modifying or eliminating those items which negatively affect its psychometric properties. It will also serve as the base for the design of a training program related to ADHD aimed at infant and primary school teachers in the Autonomous Community of the Basque Country.

## 5. References

American Psychiatric Association (APA, 2000). *Diagnostic and statistical manual of mental disorders* (4th. ed.). Author text revision, Washington.

Arcia, E., Frank, R., Sánchez-LaCay, A., & Fernández, M. C. (2000). Teacher understanding of ADHD as reflected in attributions and classroom strategies. *Journal of Attention Disorders, 4*(2), 91-101.

Bandura, A. (2001). Social cognitive theory: An agentic perspective. *Annual Review of Psychology, 52*, 1-26.

Barbaresi, W. J., & Olsen, R. D. (1998). An ADHD Educational Intervention for Elementary Schoolteachers: A pilot Study. *Developmental and Behavioral Pediatrics, 19*(2), 94-100.

Barbero, M. I. (2003). Psicometría. In M. I. Barbero (Coord.), E. Vila, & J. C. Suárez, *Principios básicos para la construcción de instrumentos de medición psicológica* (pp. 79-135). UNED, Madrid.

Barkley, R. A. (1998). *Attention deficit hyperactivity disorder: A handbook for diagnosis and treatment.* (2nd. ed.). Guilford Press, New York.

Barkley, R. A. (1999). *Niños hiperactivos: Cómo comprender y atender sus necesidades especiales.* Paidós, Barcelona.

Barkley, R. A. (2004). Adolescents with attention-deficit/hyperactivity disorder: An overview of empirically based treatments. *Journal of Psychiatric Practice, 10*(1), 39-56.

Barkley, R. A. (2005). Prólogo. In I. Moreno, *El niño hiperactivo* (pp. 13). Pirámide, Madrid.

Barkley, R. A. (2006). *Attention deficit hyperactivity disorder: A handbook for diagnosis and treatment.* (3rd. ed.). Guilford Press, New York.

Beck, C. T., & Gable, R. K. (2001). Ensuring Content Validity: An Ilustration of the Process. *Journal of Nursing Measurement, 9*(2), 201-215.

Bekle, B. (2004). Knowledge and attitudes about Attention-Deficit Hyperactivity Disorder (ADHD): A comparison between practicing teachers and undergraduate education students. *Journal of Attention Disorders, 7*(3), 2004.

Brown, T. E. (2003). *Trastornos por déficit de atención y comorbilidades en niños, adolescentes y adultos.* Masson, Barcelona.

Bussing, R., Gary, F. A., Leon, C. E., & Garvan, C. W. (2002). General Classroom Teachers' Information and Perceptions of Attention Deficit Hyperactivity Disorder. *Behavioral Disorders, 27*(4), 327-339.

Cooper, P., & Bilton, K. M. (2002). *Attention Deficit/Hyperactivity Disorder: A practical guide for teachers*. David Fulton, London.

Ghanizadeh, A., Bahredar, M. J., & Moeini, S. R. (2006). Knowledge and attitudes towards attention deficit hyperactivity disorder among elementary school teachers. *Patient Education and Counselling, 63*, 84-88.

González-Acosta, E. (2006). *Trastorno por Déficit de Atención e Hiperactividad en el salón de clases*. (Doctoral Thesis, The Complutense University of Madrid, 2006). Available from: http://www.ucm.es/BUCM/tesis/fsl/ucm-t29215.pdf

Graczyk, P. A., Atkins, M. S., Jackon, M. M., Letendre, J. A., Kim-Cohen, J., Baumann, B. L., & McCoy, J. (2005). Urban Educators' Perceptions of Interventions for Students with Attention Deficit Hyperactivity Disorder: A Preliminary Investigation. *Behavioral Disorders, 30*(2), 95-104.

Guerra, F. R. (2010). *Teacher knowledge of Attention Deficit Hyperactivity Disorder among middle school students in South Texas*. (Doctoral Thesis, University of Texas, 2009). Available from: http://gradworks.umi.com/34/00/3400334.html

Haynes, S. N., Richard, D. C. S., & Kubany, E. S. (1995). Conent Validity in Psichological Assessment: A Functional Approach to Concepts and Methods. *Psychological Assessment, 7*(3), 238-247.

Hernández, R., Fernández-Collado, C., & Baptista, P. (2006). *Metodología de la investigación*. (4th. ed.). McGraw-Hill, Mexico.

Jarque, S., & Tárraga, R. (2008). Adaptación y validación de la escala de conocimientos sobre el Trastorno por Déficit de Atención con Hiperactividad. *Proceedings of Internacional Congress of Virtual Education*, ISBN 978-84-936132-4-2, Palma de Mallorca, april 2008.

Jarque, S., Tárraga, R., & Miranda, A. (2007). Conocimientos, concepciones erróneas y lagunas de los maestros sobre el trastorno por déficit de atención con hiperactividad. *Psicothema, 19*(4), 585-590.

Jerome, L., Gordon, M., & Hustler, P. (1994). A comparison of American and Canadian Teachers' Knowledge and Attitudes Towards Attention Deficit Hyperactivity Disorder (ADHD). *Canadian Journal of Psychiatry, 39*, 563-567.

Jones, H. A., & Chronis-Tuscano, A. (2008). Efficacy of teacher in-service training for Attention-Deficit/Hyperactivity Disorder. *Psychology in the Schools, 45*(10), 918-929.

Kos, J., Richdale, A. L., & Hay, D. A. (2006). Children with Attention deficit Hyperactivity Disorder and their Teachers: A review of the literature. *International Journal of Disability, Development and Education, 53*(2), 147-160.

Kos, J. M., Richdale, A. L., & Jackson, M. S. (2004). Knowledge about Attention-Deficit/Hyperactivity Disorder: A comparison of in-service and preservice teachers. *Psychology in the Schools, 41*(5), 517-526.

Lynn, M. R. (1986). Determination and Quantification Of Content Validity. *Nursing Research, 35*(6), 382-385.

Martínez Arias, M. R., Hernández Lloreda, M. J., & Hernández Lloreda, M. V. (2006). *Psicometría*. Alianza Editorial, Madrid.

McGartland, D., Berg-Weger, M., Tebb, S. S., Lee, E. S., & Rauch, S. (2003). Objectifying content validity: Conducting a content validity study in social work research. *Social Work Research, 27*(2), 94-104.

McKenzie, J. F., Wood, M. L., Kotecki, J. E., Clark, J. K., & Brey, R. A. (1999). Establishing Content Validity: Using Qualitative and Quantitative Steps. *American Journal of Health Behavior, 23*(4), 311-318.

Miranda, A., Jarque, S., & Soriano, M. (1999). Trastorno de hiperactividad con déficit de atención: polémicas actuales acerca de su definición, epidemiología, bases etiológicas y aproximaciones a la intervención. *Revista de neurología, 28*(2), 182-188.

Miranda, A., Roselló, B., & Soriano, M. (1998). *Estudiantes con deficiencias atencionales.* Promolibro, Valencia.

Moreno, I. (2008a). *Hiperactividad infantil: Guía de actuación.* Pirámide, Madrid.

Moreno, I. (2008b). Terapia psicológica con niños y adolescentes: estudio de casos clínicos. In F. X. Méndez, J. P. Espada, & M. Orgilés (Coord), *Tratamiento psicológico de un caso de déficit de atención e hiperactividad* (pp. 323-341). Pirámide, Madrid.

Moreno, R., Martínez, R. J., & Muñiz, J. (2004). Software, instrumentación y metodología. *Psicothema, 16*(3), 490-497.

Moreno, R., Martínez, R. J., & Muñiz, J. (2006). New Guidelines for Developing Multiple-Choice Items. *Methodology, 2*(2), 57-64.

Moreno, I., & Servera, M. (2002). Intervención en los trastornos del comportamiento infantil. In M. Servera (Coord.), *Los trastornos por déficit de atención con hiperactividad* (pp. 217-253). Pirámide, Madrid.

Muñiz, J. (2001). *Teoría clásica de los tests.* (7th. ed.). Pirámide, Madrid.

Niznik, M. E. (2005). *An exploratory study of the implementation and teacher outcomes of a program to train elementary educators about ADHD in the schools.* (Doctoral Thesis, University of Texas, 2004). Available from: http://gradworks.umi.com/31/43/3143438.html

Ohan, J. L., Cormier, N., Hepp, S. L., Visser, T. A. V., & Strain, M. C. (2008). Does Knowledge About Attention-Deficit/Hyperactivity Disorder Impact Teachers' Reported Behaviors and Perceptions?. *School Psychology Quarterly, 23*, 436-449.

Parellada, M. (Coord.). (2009). *Trastorno por déficit de atención e hiperactividad: De la infancia a la edad adulta.* Alianza, Madrid.

Pfiffner, L. J. (1999). Potenciar la educación en la escuela y en casa: métodos para el éxito desde párvulos hasta el bachillerato. In R. Barkley (Ed.), *Niños hiperactivos. Cómo comprender y atender sus necesidades especiales* (pp. 245-263). Paidós, Barcelona.

Polit, D. F., & Beck, C. T. (2006). The Content Validity Index: Are You Sure You Know What's Being Reported? Critique and Recommendations. *Research in Nursing & Health, 29*, 489-497.

Sherman, J., Rasmussen, C., & Baydala, L. (2008). The impact of teacher factors on archievement and behavioural outcomes of children with Attention Deficit/Hyperactivity Disorder (ADHD): a review of the literature. *Educational Research, 50*(4), 347-360.

Sciutto, M. J., Terjesen, M. D., & Bender, A. S. (2000). Teachers' knowledge and misperceptions of Attention-Deficit/Hyperactivity Disorder. *Psychology in the Schools, 37*(2), 2000.

Sireci, S. (1998). Gathering and Analyzing Content Validity Data. *Educational Assessment, 5*(4), 299-321.

Snider, V. E., Busch, T., & Arrowood, L. (2003). Teacher Knowledge of Stimulant Medication and ADHD. *Remedial and Special Education, 24*(1), 46-56.

Stacey, M. A. (2003). *Attention-Deficit/Hyperactivity Disorder: General Education Elementary School Teachers' Knowledge, Training, and Ratings of Acceptability of Interventions.* (Doctoral Thesis, University of South Florida, 2003). Available from: http://etd.fcla.edu/SF/SFE0000084/Thesis.pdf

Syed, E. U., & Hussein, S. A. (2009). Increase in Teachers' Knowledge About ADHD After a Week-Long Training Program. *Journal of Attention Disorders, 13*(4), 420-423.

Vereb, R. L., & DiPerna, J. (2004). Teachers' Knowledge of ADHD, Treatments for ADHD, and Treatment Acceptability: An Initial Investigation. *School Psychology Review, 33*(3), 421-428.

West, J., Taylor, M., Houghton, S., & Hudyma, S. (2005). A Comparison of Teachers' and Parents' Knowledge and Beliefs About Attention-Deficit/Hyperactivity Disorder (ADHD). *School Psychology International, 26*(2), 192-208.

White, S. W., Sukhodolsky, D. G., Rains, A. L., Foster, D., McGuire, J. F., & Scahill, L. (2011). Elementary School Teachers' Knowledge of Tourette Syndrome, Obsessive-Compulsive Disorder, & Attention-Deficit/Hyperactivity Disorder: Effects of Teacher Training. *Journal of Developmental Pshysical Disabilities, 23*(5), 5-14.

# 10

# ADHD Symptomatology, Academic Dishonesty, and the Use of ADHD Stimulant Medications Without a Prescription

Kelly Custode and Jill M. Norvilitis
*Buffalo State College,*
*USA*

## 1. Introduction

Concern about the misuse of stimulant medications for Attention Deficit Hyperactivity Disorder (ADHD) among college students has been increasing in recent years, with numerous studies reporting widespread illicit use of the medications. Students who use stimulant medications without a prescription often report that they use stimulants for academic reasons, although recent research suggests that many students may be self-medicating undiagnosed symptoms of ADHD (Peterkin, Crone, Sheridan & Wise, 2011). Although researchers have begun to examine the risk factors and correlates of illicit stimulant use, questions remain about how students who use and do not use the medications view stimulant use and the degree of academic distress among those using stimulants.

ADHD has been characterized as an inability to sustain attention, impulsivity, and hyperactivity (Weyandt & DuPaul, 2008). Affecting 3-7% of the school-aged population, it was previously thought to be limited to childhood. However, it is now recognized that ADHD symptoms frequently continue into adolescence and adulthood. For instance, Biederman, Mick and Farone (2000), studied 19-year-old boys who had been diagnosed with ADHD as children and found that as many as 90% continued to exhibit at least sub-threshold levels of ADHD symptoms. In these cases, symptoms tend to present themselves differently. For example, the hyperactive motor activity expressed in a school aged child is more likely to manifest itself as mental or emotional restlessness in an adult.

ADHD is often treated by a combination of behavior modification techniques and stimulant medications (Svetlov, Kobeissy, & Gold, 2007). Stimulant medications include methylphenidate (MPH) and amphetamines which are dispensed under the brand names Adderall, Focalin, Concerta, or Ritalin. These medications work on the dopamine and noradrenaline systems in the brain, and can help to reduce fatigue, increase attention and alertness, and suppress appetite. MPH has been shown to work by binding to the dopamine transporter and inhibiting dopamine uptake. Amphetamines, on the other hand, penetrate the cell membrane through transport and diffusion in order to interact with the vesicular monoamine transporter-2. Vesicular dopamine is displaced and a newly synthesized intraneuronal monoamine is activated by means of reverse transport. This can result in a rapid build-up of synaptic monoamine, which does not occur in the use of MPH; thus increasing the abuse potential in amphetamines over MPH (Svetlov et al., 2007), although abuse remains a possibility with MPH.

The abuse potential of these medications increases as people begin to share or misuse their prescription, or obtain such medications without a prescription. When prescribing a stimulant medication, physicians begin with a low dose, and slowly increase to therapeutic levels in order to reduce the risk of addiction and side effects. Along with the benefits, stimulant medication use also comes with potential side effects; such as headaches, stomach aches, sleep problems, and tics (Hall, Irwin, Bowman, Frankenberger & Jewett, 2005). These side effects can be minimized with prescriber supervision, however those using without a prescription do not have the benefits of these precautions (Svetlov et al., 2007).

Stimulant medications are known as "universal performance enhancers", because they increase attentiveness and wakefulness in all populations; child or adult, ADHD or not (Svetlov et al., 2007). However, it is important to note that multiple studies have shown that ADHD medications do not improve academic ability or cognitive skills, such as adaptability, planning, and acquisition of new information (For review, see Advokat, 2010). Despite the lack of evidence for cognitive enhancement, there is a widespread belief, even among those who do not use stimulant medications at all, that these medications will help improve academic achievement (Advokat, Guidry, & Martino, 2008).

The National Survey on Drug Use and Health estimates that roughly 14.2% of full time college students between the ages of 18 and 22 used prescription medications for non-medical reasons at least once between 2002 and 2004 (Ford & Schroeder, 2009). More recent studies indicate higher rates with up to 35% of surveyed undergraduate students reporting the use of stimulant medications without a prescription (DeSantis, Webb, & Noar, 2008; Judson & Langdon, 2009; Wilens et al., 2008). The reasons for illicit use of these medications include experimentation, getting high (Ford & Schroeder, 2009), improving academic performance or weight loss (Advokat, 2009). The use of these medications for recreational purposes is more prevalent between the ages of 18 and 25 than any other age group (Hall et al., 2005). Additionally, some use stimulant medications without a prescription as a means to alleviate some sort of distress (Kadison, 2005). Many believe that stimulant medications are a safe way to maximize performance with minimal risk. With the way these drugs are marketed, many are led to believe that stimulants are a "magic bullet for complex problems", thus potentially influencing healthy people to seek out such medication use (Kadison, 2005).

Those who have a tendency to misuse stimulant medications may choose to take that route instead of "street substances", because they are perceived to be safer (Ford & Schroeder, 2009). This is, in part, due to the fact that these medications are considered to be "pure" because they have a known chemical composition. These medications are easier to obtain, and there is less likelihood of arrest. Stimulants are also generally perceived as more socially acceptable (Ford & Schroeder, 2009). Those looking to obtain stimulant medications without a prescription have discovered the ability to do so by buying them off of their peers, or over the internet (Kadison, 2005). Further, the ability to obtain a prescription despite the lack of necessity facilitates the misuse of these medications. Those looking to obtain a prescription for stimulant medications know which symptoms to report to their physicians (Kadison,

2005). In recent studies of adults being assessed for ADHD, between one-fifth and one-half of all patients are believed to have exaggerated their symptoms (Marshall et al., 2010; Sullivan, May, & Galbally, 2007). Given the ease with which symptoms are faked, prescriptions for stimulant medications can be easy to obtain.

One explanation for illicit medication use is the theory of planned behavior (Judson & Langdon, 2009). The theory of planned behavior states that the attitudes, beliefs about social norms, and perceived control work together in order to create intentions predictive of health related behavior. For example, if a student were to believe that the illicit use of stimulant medications were safe and ethical, while simultaneously believing that others feel the same way and the drug will improve his/her ability to control behavior, this student would be likely to use the medication illicitly. In contrast, those who believe such use is wrong, and that others also believe it is wrong are much less likely to use (Jusdon & Langdon, 2009).

Another theory behind the misuse of stimulant medications is the general strain theory (Ford & Schroeder, 2009). The general strain theory states that there are three elements that contribute to the cause of stress: the failure to achieve positively valued goals, the removal of positively valued stimuli, and the addition of noxious stimuli. From an academic standpoint, good grades and overall academic success would count as a positively valued goal. Those who are unable to meet this goal, regardless of the amount of effort put forth, are likely to experience strain. Dropping of grades or a loss of financial aid, in addition to negative encounters with faculty members can be seen as a removal of positively valued stimuli. Experiencing verbal or physical abuse from peers, and harsh reactions or judgment from faculty would fall under the addition of noxious stimuli. With a collaboration of all three elements, a student is more likely to experience negative affective states. Therefore, academic strain can be related to depression, which in turn is related to the non-medical use of prescription stimulants. In this case, the medication is used as an academic tool by means of providing the ability to stay up later and focus, while also working as an emotional tool in order to escape the stress of daily life (Ford & Schroeder, 2009).

The primary self-reported motivation for stimulant misuse in college is academic (Judson & Langdon, 2009). Reports of misuse seem to be higher in the northeastern part of the United States, in schools where admission requirements are more stringent. Students have also reported discontinuing the use of stimulant medications with a prescription in high school, yet finding it necessary to start up again in college. Additionally, some students admit to stockpiling their medications, and using more than directed in times of high academic demand (Judson & Langdon, 2009).

The choice to use stimulants, then, may be precipitated by stress, fear of failure, and academic struggle, leading students to self-medicate. Students who use stimulant medications without a prescription may also be struggling with undiagnosed symptoms of ADHD. Indeed, a recent study found that 71% of stimulant misusers had a positive screening for adult ADHD (Peterkin et al., 2010), indicating that they had enough symptoms to seek medical attention for an official diagnosis and the use of medication, though not the path to that use, may not necessarily be inappropriate.

ADHD is commonly known for its relation with academic impairment (Weyandt & DuPaul, 2008). In fact, even students with subclinical ADHD struggle academically. These students tend to have greater difficulty weeding out useless information, and honing in on relevant aspects of a lecture (Norwalk, Norvilitis, & MacLean, 2009). Although subclinical ADHD students have been found to commit less time to studying and achieving goals, those with a clinical diagnosis of ADHD demonstrate further academic impairment. Those with the hyperactive-type of ADHD report less education, lower grades, and more failed courses than students without ADHD by adulthood (Advokat, 2009). Although college students with ADHD tend to have better study skills and a greater record of academic success than those with ADHD who do not attend college, they still function below the level of college students without the disorder (Weyandt & DuPaul, 2008). ADHD college students have lower grade point averages, report more academic problems, and are more likely to end up on academic probation than the rest of the college population. In addition, these students are also less likely to attend classes and graduate from the university (Weyandt & DuPaul, 2008).

Regardless of whether students meet criteria for ADHD or not, the use of stimulant medications without a prescription may indicate that students are struggling with adjustment to college. Students with clear motivations to attend college (including those attending for personal or financial gain) have better study habits, higher grade point averages, and a stronger ability to make decisions than students attending college out of expectation or default (Phinney, Dennis, & Osorio, 2006). This is due to a level of "motivational readiness" that those attending for ambiguous reasons seem to lack (Norvilitis, Reid, Ling, & Chen, in press). Motivations are likely to influence how students engage and respond in school settings (Phinney et al., 2006). Students who are driven by the expectations of others, who are attending college out of obligation, or those attending simply to avoid a less desirable option are not as likely to put forth as much effort, and therefore succeed as students attending college with a meaningful purpose. The value that one has in education also influences the motivations and efforts of the college student (Phinney et al., 2006). As one might expect, research indicates that those with ADHD struggle with motivation to attend college (Norvilitis et al., in press).

In addition, those struggling with these types of motivation and academic concerns are likely to be more accepting of other cheating behaviors. These behaviors include copying homework, plagiarizing, and any other type of academic dishonesty or forgery (Engler, Landau, & Epstein, 2008). Although it is estimated that between 66% and 75% of college students participate in cheating behaviors, students are less likely to do so when they are able to display their own level of competence within a particular subject. Another predictor of cheating behavior is the students' perceptions of their peers' attitudes towards the concept of cheating itself. Those who believe their friends and classmates partake in cheating behaviors are more likely to cheat as well (Engler et al.,2008). Students who take stimulant medications without a prescription with the intent to better their grades are engaging in a form of academic dishonesty. Regardless of whether or not the medications actually improve cognitive functioning, student who use stimulants without a prescription often do so with the specific intent of improving their grades. The motivation to participate in this type of cheating behavior can be due to a lack of mastery in a subject, or a belief that several others within the school do so as well.

Thus, the present study sought to examine ADHD symptomatology, motivations to attend college, and perceptions of cheating among those who do and do not use ADHD stimulant medication without a prescription. It was expected that the study would replicate prior work indicating that those who use stimulant medication without a prescription would be more likely to have higher levels of ADHD symptomatology. It was further expected that those who use stimulant medication without a prescription report more default and less career or personal motives to attend college than those who do not report such use. It was also expected that stimulant users would report more acceptance of cheating behaviors and more cheating. In addition, the study examined beliefs about whether stimulant use is a cheating behavior and beliefs about the prevalence of side effects. It was expected that those who use stimulants would underestimate the side effects and that students would view stimulant use as cheating.

## 2. Method

### 2.1 Participants

The 184 participants of this study included 36 male (19.7%) and 147 female (80.3%) college students, with one student failing to report gender. Of these, 42 (22.8%) were between the ages of 18-19, 79 (42.9%) were 20-21, 35 (19%) were 22-23, 7 (3.8%) were 24-25, and 21 (11.4%) were 26 and older. Further, 16 (8.7%) were in their freshman year of school, 43 (23.4%) were sophomores, 76 (41.3%) were juniors, 46 (25%) were seniors, and 3 (1.6%) were fifth-year seniors or beyond. In addition, 21 (11.5%) were natural science majors, 117 (63.6%) were in social science, 16 (8.7%) were in arts and humanities, 4 (2.2%) education, and 25 (13.7%) were majoring in applied fields, with one student failing to report a major.

The majority of the participants were white ($n$ = 125, 68.7 %), 43 (23.6%) were African-American, 10 (5.4%) Hispanic, 2 (1.1%) Asian, and 2 (1.1%) Native American, with 2 students failing to report ethnicity. Participants had a range of grade point averages (GPA), with 33 (18.0%) having a GPA from 3.51-4.0, 49 (26.6%) with 3.01-3.5, 59 (32.2%) with 2.56-3.0, 33 (18.0%) with 2.01-2.5, and 9 (4.9%) with a GPA of 2.0 or below, with one student failing to report a GPA.

Five (2.7 %) participants reported that they were diagnosed with ADHD in the past but no longer met criteria for the disorder. Ten (5.4%) reported a current diagnosis of ADHD.

### 2.2 Materials and procedure

Following the completion of written informed consent, participants were asked to complete a survey consisting of demographic information and several scales to assess several factors:

**Attention Deficit Hyperactivity Disorder**. ADHD was assessed using the Current Symptoms Scale (CSS; Barkley & Murphy, 1998). The CSS is a self-report measure comprised of 18 symptoms, with 9 items each comprising the inattentive and hyperactivity subscales. All items are scored on a scale ranging from 1 (*Never or Rarely*) to 4 (*Very Often*). Higher scores indicate more ADHD symptoms. The authors recommend a cut-off score of 1.5 standard deviations above the mean to suggest possible ADHD (Barkley & Murphy, 1998). Test-retest reliability of this measure is good ($r$ = .82) and internal consistency is adequate (Cronbach's alpha .63 to .75; Aycicegi, Dinn, & Harris, 2003). In the present study, internal consistency was .84 for the inattentive scale and .79 for the hyperactivity scale.

In addition, participants were asked if they have ever been diagnosed with ADHD but no longer meet criteria for the disorder, if they are presently diagnosed with ADHD, and if they believe that they should be screened for ADHD.

| Item | Mean (SD) |
|---|---|
| It is ok to take stimulant medications that are prescribed while studying the night before a test. | 2.72 (1.22) |
| It is okay to take memory enhancing vitamins throughout the semester to try to get ahead in school. | 2.79 (1.11) |
| It is considered cheating to work in a group and not contribute. (R) | 3.23 (1.16) |
| It is considered cheating to lie about family circumstances in order to get an extension on an assignment. (R) | 3.36 (1.25) |
| It is okay to copy another person's work, as long as I cite it. | 3.41 (1.23) |
| It is okay to take a sedative before a test to calm my nerves. | 3.51 (1.07) |
| It is not okay to copy another person's work under any circumstances. (R) | 3.54 (1.33) |
| If I ran out of time to do my homework, it is okay to copy a friend's homework to get an A, as long as I understand the material. | 3.57 (1.20) |
| It is okay to copy a friend's homework, as long as I return the favor. | 3.75 (1.15) |
| It is okay to take stimulant medications that are not prescribed while studying the night before a test. | 3.84 (1.07) |
| I am cheating myself if I do not take the time to truly learn the Information provided in class. (R) | 3.94 (1.08) |
| If I do not understand the material, it is okay to copy a friend's homework so I can get an A. | 3.95 (1.04) |
| It is okay to copy another person's work, as long as it does not happen often. | 4.13 (1.00) |
| It is okay to copy another person's work and not cite it, as long as it is not the entire assignment. | 4.22 (0.93) |
| It is okay copy another person's work, as long as nobody else is aware of it. | 4.25 (0.88) |
| If I had trouble sleeping the night before, it is okay to copy a friend's test. | 4.28 (0.86) |

*Note.* Means for (R) items have been reverse scored to be consistent with the other items. Lower scores indicate greater acceptance of the behavior.

Table 1. Acceptability of cheating scale

**Motivation to Attend College.** Motivation to was assessed using the Student Motivation for Attending University-Revised (SMAU-R; Phinney et al., 2006). The SMAU-R is comprised of 7 subscales designed to tap different motivations students may have for attending college. These include Career/Personal (Sample item: "To achieve personal success"; 10 items); Humanitarian ("To contribute to the welfare of others"; 4 items); Default ("There are few other options."; 6 items); Expectation ("Parents/family would be very disappointed."; 5 items); Prove Worth ("To prove to others that I can succeed in college."; 3 items); Encouragement ("I was encouraged by a mentor or role model."; 3 items) and Help Family ("It would allow me to help parents/family financially"; 2 items). The authors reported good to strong reliabilities, ranging from .70 (Encouragement) to .87 (Help Family). Items were scored on a 1 (*Strongly Agree*) to 5 (*Strongly Disagree*) scale. Thus, higher scores indicate lower levels of agreement with each motivation. Subscale composite scores are averages of the items in the factor, to allow comparisons between the subscales. In the present study, reliability varied between subscales, but was generally good (Career/Personal: α = .76; Humanitarian: α = .80; Default α = .71; Expectation: α = .83; Prove Worth: α = .85; Encouragement: α = .75; Help Family: α = .78).

**Acceptability of Cheating.** Sixteen items were created for the present study to examine acceptability of various cheating behaviors and attitudes toward various behaviors (See Table 1). All items were rated on a 5-point scale from 1 (*Strongly Agree)* to 5 (*Strongly Disagree).* Internal consistency for the new scale was good (α = .85). Higher scores indicate less approval of cheating. Although the total score is intended to be used as an indication of overall acceptability of cheating, it is also interesting to look at the mean scores for each item to examine the relative acceptability of individual behaviors and attitudes.

**Perceptions of Psychostimulants.** Nineteen items were created to examine perceptions of psychostimulant use. Each item was rated on a 5-point scale from 1 (*Strongly Agree)* to 5 (*Strongly Disagree).* Items were grouped into three factors: beliefs about stimulant use for ADHD, beliefs about stimulant use without ADHD, and attitudes toward the acceptability of stimulant use. Principal components factor analysis with varimax rotation indicated support for the three factor model, with two items failing to load on any factor, creating a 17-item final scale (See Table 2). Internal consistency of the three subscales was good (Beliefs about Stimulant Use for ADHD, α = .85; Beliefs about Stimulant Use without ADHD, α = .91, and Attitudes toward the Acceptability of Stimulant Use, α = .78).

| Item |
|---|
| Factor 1: Beliefs about Stimulants without a Prescription (Initial Eigenvalue 6.49) |
| Stimulant medications help those without ADHD to improve their grades. |
| Stimulant medications help those without ADHD sustain attention. |
| Stimulant medications help those without ADHD to complete tasks on time. |
| Stimulant medications help those without ADHD to make new friends. |
| Stimulant medications help those without ADHD to think more critically. |
| Stimulant medications help those without ADHD to learn new material faster. |
| Stimulant medications work better on those without ADHD, because they are less impaired. |
| Factor 2: Beliefs about Stimulants for ADHD (Initial Eigenvalue 2.83) |
| Stimulant medications help those with ADHD improve their grades. |
| Stimulant medications help those with ADHD sustain attention. |
| Stimulant medications help those with ADHD complete tasks on time. |
| Stimulant medications help those with ADHD to think more critically. |
| Stimulant medications help those with ADHD to learn new material faster. |
| Factor 3: Attitudes about the Acceptability of Stimulant Use (Initial Eigenvalue 1.87) |
| Stimulant medications work the same on those with and without ADHD. |
| It is safe to take stimulant medications without a prescription, as long as the dose is what a doctor would prescribe if I were to have a prescription. |
| It is safe to take stimulant medications without a prescription, as long as I express symptoms of ADHD. |
| It is safe to take stimulant medications because they are harmless. |
| It is socially acceptable to take stimulant medications. |

Table 2. Perceptions of Stimulant Use

**Stimulant Use.** A single item asked if participants currently take stimulant medication without a prescription or not. In addition, five items asked how often participants have taken stimulant medications in the past six months to improve school work, to get high, to get more energy or stay up late, to enhance social interactions, and to lose weight. These five items were summed to create a combined stimulant use score. Participants were also asked which of these reasons was their primary reason for the use of stimulant medications without a prescription.

In addition, participants were asked if they have a prescription for psychostimulants and if they have friends who take stimulant medication with or without a prescription.

Finally, to assess knowledge and experience of side effects, participants completed two scales. The first asked if it is possible to experience various symptoms on a five-point scale from never to very often and the second asked if participants had experienced these symptoms on a three-point scale (never, once, more than once). Ten side effects were listed, six that are actual side effects (headache, nausea, stomachache, changes in appetite, sleeplessness, and dizziness) and four that are not (sore throat, temporary numbness in arms or legs, runny nose, higher sensitivity to pain). For the knowledge of side effects, mean scores were calculated for the actual side effects and the non-side effects, allowing for the calculation of a difference score. Lower scores indicated lower levels of knowledge of actual side effects.

## 3. Results

### 3.1 Prevalence of stimulant use

A total of 18 participants (17.7%) reported currently using stimulant medications without a prescription, and 14 participants (7.8 %) reported using such medications with a prescription. In addition, 50 students (27.2%) reported using non-prescription stimulants at some point in the last six months. The primary motivation for illicit use of stimulant medications was to get high (44%), followed by staying up late (20 %), improving grades (18 %), and losing weight (18 %).

Eighty-three participants (45.6%) reported having a friend who takes stimulant medications without a prescription. Those who reported having friends who take stimulants were more likely to report use themselves [$F$ (1, 171) = 16.65, $p < .001$, $\eta^2 = .09$].

A total of 42 (24 %) participants (both prescribed and illicit users) reported experiencing at least one side effect while using stimulant medication. Side effect questions asked if the person had ever experienced the particular side effect, so it is possible that those who had used stimulants more than six months ago but not in the last six months also endorsed these items. The most commonly reported side effect was sleeplessness, with 36 (19.6%) participants reporting experiencing this at least once following use.

### 3.2 Attitudes toward and knowledge of stimulant use

Students distinguished between the use of stimulants for those with ADHD and stimulants for those without ADHD. Beliefs about stimulant use for ADHD were very modestly related to attitudes toward the acceptability of stimulant use ($r = .15$, $p = .04$), whereas beliefs about stimulant use without ADHD were much more strongly related to attitudes toward the acceptability of stimulant use ($r = .41$, $p < .001$).

As might be expected, those who reported more use of stimulants had more positive beliefs about stimulants for ADHD ($r = .18, p = .02$), stimulant use without ADHD ($r = .22, p = .004$, and the overall acceptability of stimulants ($r = .32, p < .001$). Knowledge of the side effects of stimulant use was related to actually having experienced side effects ($r = .18, p = .02$), but was not related to beliefs about stimulant use for ADHD ($r = -.10, p =.19$), beliefs about stimulant use without ADHD ($r = -.05, p = .48$), or the acceptability of stimulant use ($r = .04, p = .56$).

### 3.3 Relationship between stimulant use, attitudes toward cheating, and the motivation to attend college

There was a relationship between acceptance of cheating and illicit stimulant use ($r = .40, p = .001$), such that those who used stimulants were more accepting of cheating. Illicit stimulant use was also related to students only attending college to prove someone wrong ($r = .19, p = .01$), or because they did not know what else to do with their lives ($r = .16, p = .04$). Stimulant use was unrelated to the remaining motivations to attend college (Career $r = -.08$, $p = .28$, Help Family $r = -.05$, $p = .55$, Encouragement from Others $r = -.04$, $p = .61$, Expectation of Others $r = .02$, $p = .83$, and Humanitarian reasons $r = .01$, $p = .90$).

### 3.4 Stimulant use, attitudes toward stimulant use, and ADHD symptomatology

Illicit stimulant medication use was related to self-reported inattentive ($r = .21, p = .007$) and hyperactive symptoms ($r = .29, p < .001$). Further, those who reported a belief that they should be assessed for ADHD were also more likely to report stimulant use [$F$ (1, 167) = 27.21, $p < .001$, $\eta^2 = .14$].

## 4. Discussion

The results of the present study support prior research indicating that the nonprescription use of stimulant medications is widespread on campus (Ford & Schroeder, 2009; Judson & Langdon, 2009). Although the primary reported reason is to get high, a number of students reported use to improve grades or to stay up late. One possible reason for needing assistance to stay up late is to get the necessary energy to complete school work.

The use of stimulants without medical supervision is unsafe for many reasons and there are many well-established side effects. However, the present study identified a number of other concerns associated with their use. Notably, those who reported use were also more likely to be accepting of academic dishonesty. They also report attending college for reasons other than the desire to truly learn. These are the students who are attending for reasons of default or expectation. That is, they are in college because they were not sure what else they could be doing or because someone expected it of them. This combination of motivations to attend college and acceptance of cheating suggests that these students may be floundering in college without direction. Without an intrinsic goal, attending college becomes much more tedious and overwhelming than originally anticipated. This general sense of feeling lost within the college community not only contributes to academic strain, but may lead to distress as well. Indeed, many students who use stimulants without a prescription believe that they should be tested for ADHD. In fact, as also found in the Peterkin and colleagues study (2010), they report more ADHD symptoms than those who do not use stimulants without a prescription, supporting the idea that this is a group that is struggling.

In addition, these students tend to be more likely to have at least one friend who takes stimulant medications without a prescription. That is, those who are accepting of stimulant misuse and cheating behaviors tend to travel in groups with those who maintain similar beliefs. It is possible, then, that their friends are similarly lost in college.

It is difficult to know which came first: the default or expectation motives to attend college, the higher levels of self-reported inattention and hyperactivity, peer use of stimulants or personal stimulant use. What is clear is that simply addressing non-prescription stimulant use will likely not be sufficient. Colleges must address the lack of direction and ADHD symptomatology if they wish to tackle the problem of stimulant use on campus. Although not examined in this study, it is possible that addressing these issues would improve academic performance and increase retention.

On a more positive note, although many students report using stimulants without a prescription, their use is not widely condoned. When asked about the acceptability of various forms of cheating, the non-prescription use of stimulants is about as acceptable as copying homework, which is in the middle of the hierarchy. Thus, the use of stimulants to improve grades is considered more acceptable than cheating on a test or final project, which sits at the top of the cheating hierarchy, but less acceptable than lying about family circumstances to get an extension on an assignment.

Despite the intriguing results and potential implications of this research, there are limitations as well. First, there were not enough male participants to meaningfully investigate any gender differences. Future research should examine this issue. Second, there are questions about the reliability of self-report measures. It is possible that some participants may have been concerned about reporting symptoms or use, despite the anonymous nature of the questionnaire. Given that concern, it is possible that these issues were underreported in the present study.

Overall, however, this study has confirmed prior research on the prevalence of use and expanded that research to examine academic concerns among those who use stimulants without a prescription. Many questions persist, however. Future research should investigate whether any interventions would prove useful in reducing stimulant use and improving motivation to attend college.

## 5. References

Advokat, C. (2009). What exactly are the benefits of stimulants for ADHD? *Journal of Attention Disorders, 12*, 495-497.

Advokat, C. (2010). What are the cognitive effects of stimulant medications? Emphasis on adults with attention deficit hyperactivity disorder. *Neuroscience and Biobehavioral Reviews, 34*, 1256-1266.

Advokat, C., Guidry, D., & Martino, L. (2008). Licit and illicit use of medications for attention deficit hyperactivity disorder in undergraduate college students. *Journal of American College Health, 56*, 601-606.

Barkley, R.A. & Murphy, K.R. (1998). *Attention deficit hyperactivity disorder: A clinical workbook* (2nd ed.). New York: Guilford Press.

Biederman, J., Mick, E., & Faraone, S.V. (2000). Age-dependent decline of symptoms of attention deficit hyperactivity disorder: Impact of remission definition and symptom type. *The American Journal of Psychiatry, 157,* 816-818.

DeSantis, A. D., Webb, E. M., & Noar, S. M. (2008). Illicit use of prescription ADHD medications on a college campus: A multimethodological approach. *Journal of American College Health, 57,* 315-323.

Engler, J. N., Landau, J. D., & Epstein, M. (2008). Keeping up with the Joneses: Students' perceptions of academically dishonest behavior. *Teaching of Psychology, 35,* 99-102. doi:10.1080/00986280801978418

Ford, J. A., & Schroeder, R. D. (2009). Academic strain and non-medical use of prescription stimulants among college students. *Deviant Behavior, 30,* 26-53.

Hall, K. M., Irwin, M. M., Bowman, K. A., Frankenberger, W., & Jewett, D. C. (2005). Illicit use of prescribed stimulant medication among college students. *Journal of American College Health, 53(4),* 167-174.

Judson, R., & Langdon, S. W. (2009). Illicit use of prescription stimulants among college students: Prescription status, motives, theory of planned behavior, knowledge, and self-diagnostic tendencies. *Psychology, Health & Medicine, 14(1),* 97-104.

Kadison, R. (2005). Getting an edge—Use of stimulants and antidepressants in college. *The New England Journal of Medicine, 353,* 1089-1091.

Marshall, P., Schroeder, R., O'Brien, J., Fischer, R., Ries, A., Blesi, B., & Barker, J. (2010). Effectiveness of symptom validity measures in identifying cognitive and behavioral symptom exaggeration in adult attention deficit hyperactivity disorder. *The Clinical Neuropsychologist, 24,* 1204-1237.

Norvilitis, J. M., Reid, H. M., Ling, S., & Chen, S. (in press). Motivation to attend college: A comparison of American and Chinese students, and the effects of ADHD symptomatology and personality upon their motivation. *Journal of the First Year Experience and Students in Transition.*

Norwalk, K., Norvilitis, J. M., & MacLean, M. G. (2009). ADHD symptomatology and its relationship to factors associated with college adjustment. *Journal of Attention Disorders, 13,* 251-258. DOI: 10.1177/1087054708320441

Peterkin, A. L., Crone, C. C., Sheridan, M. J., & Wise, T. N. (2011). Cognitive performance enhancement: Misuse or self-treatment? *Journal of Attention Disorders,15,* 263-268.

Phinney, J. S., Dennis, J., & Osorio, S. (2006). Reasons to attend college among ethnically diverse college students. *Cultural Diversity and Ethnic Minority Psychology, 12,* 347-366.

Sullivan, B., May, K., & Galbally, L. (2007). Symptom exaggeration by college adults in attention deficit hyperactivity disorder and learning disorder assessments. *Applied Neuropsychology, 23,* 521-530.

Svetlov, S. I., Kobeissy, F. H., & Gold, M. S. (2007). Performance enhancing, non-prescription use of Ritalin: A comparison with amphetamines and cocaine. *Journal of Addictive Diseases,26(4),* 1-6.

Weyandt, L. L., & DuPaul, G. J. (2008). ADHD in college students: Developmental findings. *Developmental Disabilities, 14,* 311-319.

Wilens, T. E., Adler, L. A., Adams, J., Sgambaait, S., Rotrosen, J., Sawtelle, R., Utzinger, L., & Fusillo, S. (2008). Misuse and diversion of stimulants prescribed for ADHD: A systematic review of the literature. *Journal of the American Academy of Child and Adolescent Psychiatry, 47,* 21-31.

# 11

# Understanding the Distracted and the Disinhibited: Experiences of Adolescents Diagnosed with ADHD Within the South African Context

J. Seabi* and N.A. Economou
*Department of Psychology, School of Human and Community Development, University of the Witwatersrand, South Africa*

## 1. Introduction

This chapter explores how adolescents diagnosed with Attention Deficit Hyperactivity Disorder (ADHD) make sense of their condition, how they feel about their difficulties and how they perceive themselves and their reality within the South African context. This chapter is presented against the backdrop of more than thirty years of well-established research that has largely focused on assessment, diagnosis, management and treatment of the condition. The intention of this chapter is to examine how adolescents perceive ADHD as it relate to their lives. The accounts contained in the research provide insights into the lives and minds of young people who have been diagnosed as having ADHD. Qualitative approach was employed and data was collected via semi-structured interviews. The sample comprised five adolescents, all outpatients of Chris Hani Baragwanath's Child and Adolescent Psychiatric Unit. Through a process of thematic content analysis the data was analyzed and major themes were identified and interpreted.

## 2. Rationale

Attention Deficit Hyperactivity Disorder is currently the most commonly diagnosed behavioral childhood disorder and the diagnosis is without doubt controversial. ADHD continues to receive ongoing attention from parents, members of education systems, health professionals and researchers. In the United States of America, between 30 percent and 40 percent of all children who are referred to child clinics and centers have been diagnosed with ADHD (Bird, 2002; Timimi & Taylor, 2004; Polanczk, de Lima, Horta, Biederman & Rohde, 2007). The prevalence rates in South Africa and Africa are reported to be just as high as in Western countries (Aase, Meyer, Sagvolden, 2006; Adewuya & Famuyiwa, 2007, Louw, Oswald & Perold, 2009; Seabi, 2010; Van der Westhuizen, 2010). It is thus imperative that the psychological and educational domains embrace and work with this condition.

Despite the limited studies (Meyer, Eilertsen, Sundet, Tshifularo & Sagvolden, 2004; Mahomedy, van der Westhuizen, van der Linde & Coetsee, 2007) conducted within the African continent, not much is known about experiences and knowledge of adolescents with ADHD in Africa. To exacerbate the situation, ADHD is the most prevalent child psychiatric disorder in South Africa (Mahomedy et al., 2007; Perold, Louw & Kleynhans, 2010).

To date, over thirty years of research has been invested in the understanding of ADHD. Knowledge of the condition, with particular reference to, diagnosis, management and treatment, is well-established and documented (Goldman, Genel, Bezman & Slanetz, 1998; Polanczk et al., 2007). Without this significant body of work, educationalists and health professionals would not be in a position to intervene or treat the child who presents with the disorder.

However, the time has now arrived for the attention to be shifted. Research needs to focus on how the child with ADHD experiences the condition, as well as the impact it has on his/her emotional and social well-being. Given a recent finding which demonstrated that psychotropic drugs such as Ritalin and Adderall may produce a placebo effect- not in the children, but in their teachers, parents and other concerned adults who examine them (Nauert, 2009), it seems pertinent to understand how those who take these medications experience them. There appears to be a gap in the literature especially within the South African context regarding the child's personal experience of living with a diagnosis of ADHD. In this way, those who work with the condition will gain a better understanding.

While it is common for ADHD to have been diagnosed in early childhood, it is a misconception that the adolescent outgrows the disorder. Van der Westhuizen (2010) concurs that ADHD was believed to resolve around puberty as most hyperactive children become less hyperactive around this time, but symptoms of impulsivity and inattention usually persist in adulthood. Recent studies reveal that around 30% to 70% of children experience problems related to ADHD in adulthood (Louw et al., 2009; Mahomedy et al., 2007).

Since adolescence is in itself a complicated developmental stage, it seems particularly significant to explore how adolescents with ADHD view their world. Adolescents with ADHD are more prone to engaging in high-risk behaviors such as drug use, early smoking and sexual activity than their peers without ADHD (Lambert & Hartsough, 1998). Aase et al. (2006) are in agreement that the disorder places an adolescent at a risk for school failure, criminality, substance abuse and sexual promiscuity.

Research indicates that interpersonal relationships become increasingly difficult as children with ADHD enter preadolescence (Kendall, 2000; Mercugliano, 1999). Mash and Wolfe (2005) confirm that children diagnosed with ADHD often experience interpersonal problems with family members, educators and peers. Their behavior is often unpredictable, hostile and confrontational with an inability to learn from their past mistakes. Therefore, research needs to be conducted on how these adolescents make sense of their condition, how they perceive themselves in the world, how they see the condition to have impacted on their development and their responses to interventions they have received.

Consequently, the intention of this chapter is to examine how adolescents perceive and experience ADHD. The accounts contained in the research provide insights into the lives and minds of young people who have been diagnosed as having ADHD. This chapter is intended to be useful to parents, teachers, health professionals, support groups and other researchers, to further the understanding, definition and treatment of ADHD.

### 2.1 Subtypes of ADHD

Briefly, to date, three subtypes of ADHD are currently recognized in the Diagnostic and Statistical Manual of Mental Disorders, Fourth Edition (DSM-IV), namely, Predominantly Inattentive Type, Predominantly Hyperactive-Impulse and a Combined Type (American Psychiatric Association, 1994). The Combined subtype is the most commonly represented subgroup, followed by the Inattentive type (Biedermann, Willens & Spencer, 2002). In Europe however, different criteria are used from the ICD-10 (International Classification of Diseases) and the condition is known as Hyperkinetic Disorder (Schroeder & Gordon, 2002). In South Africa, the aid medical insurances require ICD-10 codes when service providers submit claims and as a result, the ICD-10 is largely used than the DSM-IV.

Regarding symptomatology, Barkley (2002) defines ADHD as having a constellation of symptoms, characterized by three hallmark symptoms; inattention, hyperactivity, and impulsivity. Barkley (1998) calls these symptoms the 'holy trinity' and asserts that they give rise to problems with inhibiting, initiating or sustaining responses to tasks and adhering to rules or instructions. Furthermore, it creates significant impairment in both the home and school environment (Van der Westhuizen, 2010). ADHD like any other Disruptive Behaviour Disorders, it is regarded as an externalizing disorder, in the sense that the behavior is usually more upsetting to others than to those with ADHD themselves (Holz & Lessing, 2002).

Constructs around attention include alertness, arousal, selectivity, distractibility and sustained attention. Children and adolescents with chronic inattention experience their greatest difficulties with sustaining their attention. Inattentive symptoms include making careless mistakes, avoiding tasks, having difficulty listening to others and following instructions (Conners & Jett, 2001). Impulsivity is commonly described as the inability to delay a response or a form of gratification. Blurting out comments inappropriately or butting into conversations are manifestations of this symptom (Conners & Jett, 2001). Impulsivity is also understood as behavioral disinhibition, the inability to regulate behavior by consequences and rules (Goldstein & Goldstein, 1998). Hyperactivity signifies excessive restlessness - becoming easily aroused emotionally and physically - at an age or developmental stage that is inappropriate. Barkley (2002) purports that the nature of ADHD should not be understood as being characterized by heightened distractibility and hyperactivity, but rather by problems of behavioral disinhibition and the consistent lack of persistence to sustain attention (Goldstein & Goldstein, 1998).

### 2.2 ADHD across the lifespan

A large body of literature relates predominantly to children who have been diagnosed with ADHD. However, research emphasizes the persistence of ADHD throughout the lifespan. A substantial number of adolescents do not fully outgrow ADHD symptoms. Longitudinal follow-up studies provide strong evidence of the high percentage and continuation of

ADHD into adulthood (Amen, 2001; Biedermann et al., 2002). ADHD is now understood as a chronic lifetime disorder.

For a careful diagnosis, it is critical to understand the developmental course of ADHD. Many children with ADHD are said to be active in the womb. They are born as colicky infants who are restless sleepers, fussy eaters, are sensitive to noise and touch and have motor and attention deficits (Amen, 2001; Conners & Jett, 2001). These symptoms appear to rise relatively early in childhood with the mean age onset being between 3-4 years (Barkley, 1998). As toddlers, they are noted to become excessively active and mischievous, extremely curious, use vigorous play, have excessive temper tantrums, are demanding of parental attention and have difficulty in completing developmental tasks (Amen, 2001).

Yet, the majority of children with ADHD are recognized when they enter school. This is at the developmental stage of middle childhood. According to Barkley (1998; 2002), once they are recognized, a major social burden is placed on them. It causes a tremendous distress for many of them and their parents. Contrary to what youth with ADHD are able to do, the ability to sit still, attend, inhibit impulsive behavior, cooperate, follow through on instructions, play and interact pleasantly is crucial to a successful academic career (Conners & Jett, 2001). At home parents complain that their children with ADHD do not complete household chores and responsibilities. Parents come to recognize that they need supervision as children with ADHD become to be perceived as being immature (Rhee, Waldman, Hay & Levy, 1999).

Research suggests that ADHD does not end in childhood and the majority of children with ADHD continue to exhibit symptoms into adolescence with 40-50% carrying a diagnosis of oppositional defiant disorder and conduct disorder (Barkley, 1998; Robin, 2002; Krueger & Kendall, 2001). Longitudinal studies confirm that ADHD persist into mid-adolescence for at least 40% of the children diagnosed in childhood and 30% will continue to meet the diagnostic criteria for ADHD into adulthood (Holz & Lessing, 2002; Schroeder & Gordon, 2002). These studies revealed that changes in symptoms can decline with age. Results indicate that the frequency of parent and teacher reported hyperactivity-impulsivity symptoms decline slightly. It is understood that the symptoms of hyperactivity and impulsivity decline and inattention becomes predominant. As children get older they are diagnosed with ADHD Inattentive Type (Dupaul, Power, Anastopoulos, Reid, McGoey, et al., 1997). Conners and Jett (2001) report that although symptoms of hyperactivity may diminish, inattention and impulsivity persist and feelings of inner restlessness come to the fore.

Adolescence is a trying time and is renowned for being a complex stage of development. Its defining feature is transition, since developmentally, adolescence falls between childhood and adulthood. Adolescence is a dynamic interplay between external pressures, having to mature quickly as well as being held back or kept from independence. Above all, adolescence is marked by ambivalence (Bryant & Coleman, 1994).

Much attention in current research is turning its focus to ADHD in adolescence (Bradley-Klug & Grier, 2000; Robin, 2002). Barkley (1990, p.266) encapsulate the difficulty of ADHD adolescence; "the adolescent years of ADHD individuals may be some of the most difficult because of the increasing demands of independent, responsible conduct, as well as the emerging social and physical changes in puberty. Issues of identity, peer group acceptance, dating and courtship and physical development and appearance erupt as a source of demands and distress which the ADHD adolescent must now cope".

As this is a stage marked with physical and biological changes, development of stronger peer relationships and assertion of independence from parents, it is significant to note how adolescents with ADHD cope in their day-to-day lives. The point made earlier is that, for any adolescent, the changes are challenging. For the ADHD adolescent, the transition is exacerbated (Bradley-Klug & Grier, 2000).

Adolescents with ADHD are more likely to fail a grade, be suspended, expelled or drop out, have increased conflict, negative communication and have high risk taking behavior, bad driving habits, risky sexual behavior (Robin, 2002). Indeed studies reveal that ADHD can have a severe impact if not well managed on the learner's scholastic achievement and personal development (DuPaul & Stoner, 1994; Gross, 1997). Consistent with these studies are the findings that scholastic underachievement has far-reaching effects for learners with ADHD, such as experiencing rejection by others and self-rejection (Holz & Lessing, 2002). Furthermore, they report that ADHD learners' self-concepts and opportunities for the future are adversely affected.

The concern for ADHD teenagers is that the diagnosis in the stage of adolescence is elusive and it can be easily misdiagnosed as clinicians are less likely to consider ADHD as a possible diagnosis for behavioral problems. In addition, it is more difficult to obtain information from educators in secondary schools. Accurate diagnosis, awareness and recognition of the fact that the teenager might be suffering from ADHD are essential.

The impact of untreated ADHD on the adolescent is far more damaging for the teenager, immediate family and social context in which they live (Nahlik, 2004). Bierdermann et al. (2002) report that approximately 75% of ADHD symptoms persist into young adulthood. The study showed that the symptoms of ADHD into adulthood included inattention, distractibility, impulsivity and occupational malfunction.

Although a large body of research on ADHD has been conducted largely in developed countries in Euro-Western countries, only two studies (Kendall, Hatton, Beckett & Leo, 2003; Krueger & Kendall, 2001) were conducted in the United States which explored the experiences of adolescents with ADHD, in order to uncover how they perceive and manage ADHD. Krueger and Kendall (2001) conducted a qualitative study on a sample of 11 adolescents with age ranges of 13 to 19, who have been diagnosed with the ADHD for at least two years. The findings revealed that participants described themselves in relation to their ADHD symptoms, rather than holding an identity distinct from the disorder. Their descriptions centered on inadequacy in girls and anger and defiance in boys. Of significant interest it was that they did not view ADHD as being a disorder and the cause of many of their difficulties, but rather, the negative attributes of ADHD appeared to be incorporated into their identity.

In another study, Kendall et al. (2003) further explored children's and adolescents' accounts of their experiences with ADHD. In-depth interviews were conducted with a sample of 39 children (67% boys; 38% African-American, 33% Hispanic, 23% Caucasian) with an age range of 6 to 17 years. The results yielded six themes which were clustered as the (1) *problems*. In this cluster, the participants reported the problems that got them into trouble. These problems could be categorized as learning and cognition problems such as their slower rate of learning. They perceived themselves to be different from their peers as they were constantly distracted and confused by what they were instructed to do. They also mentioned behavioral problems which included not being able to control themselves, fighting, throwing things and yelling at their peers and teachers. In addition, they reported emotional problems in which they spoke about ADHD in terms of feeling sad, frustrated, mad, ashamed and sad. They felt that their peers laughed at them and perceived them as being stupid.

The second theme involved (2) *meaning and identity of ADHD*, whereby "they spoke about ADHD in terms of who they were, rather than the symptoms they experienced" (Kendall et al., 2003, p. 120). For them, ADHD meant being hyper- that is being out of control and unable to concentrate as well as talking too much. It is reported that despite many of them believing that they had ADHD, they did not perceive it as an illness until an authority figure told them they had ADHD. For some, they did not see ADHD as an illness but rather as part of who they were and that they were different.

The participants reported about (3) *pills*, that they took medication at some point in their life and revealed both positive and negative outcomes. Positively, they perceived medicine as something that helped them a lot, to concentrate and to control their hyperactivity. Negative outcomes included the bad taste, headaches and stomachaches, as well as the fear and shame associated with taking pills to control their behavior.

It was found that (4) *mom* was the most helpful person, who reminded them of taking the medicine. They were concerned about the effects of ADHD on their mothers and families. The other themes involved (5) *causes* in terms of what they attributed ADHD to and (6) *ethnicity* as different races described themselves differently. The implication of this study points out to the importance of and necessities of educators, parents, nurses, psychologists and others in providing a nurturing and containing environment in order to enhance their self-esteem, to feel proud and not ashamed of taking medication; thereby indirectly preventing them from being delinquents. Like many other illnesses, ADHD needs not to be seen as a personality trait but rather as an illness that can be controlled through adherence to the medication.

To the best knowledge of the authors, other than the two studies mentioned above, no empirical studies could be located nationally and internationally that investigated the perceptions of adolescents with ADHD. Therefore, the current study attempted to close the gap from an African contextual perspective.

## 3. Methods

### 3.1 Context of the study

This study was conducted in Chris Hani Baragwanath Hospital, *Child, Adolescent and Family Unit* in South Africa. The hospital situated south west of the city of Johannesburg, on the southern border of the township of Soweto. Soweto was classified as 'blacks only' which comprised the Indians, Coloureds and Africans under apartheid legislation (the Group Areas Act). The demographics have changed dramatically as a result of rapid urbanization and overcrowding due to large immigrant population of people from all corners of the country and other countries in Africa (Vogel & Holford, 1999). It is largely responsible for the health needs of approximately 5.5 million Soweto residents. Laning, Roake, Glynis and Horning (2003, p.8) describe Soweto as "one of the largest black cities on the African continent, with an eclectic mix of ethnic groups from South Africa and beyond…many holding proudly to their traditional cultural heritage while enthusiastically embracing the modern western way of life". The Child, Adolescent and Family Unit was selected as the research because of children with attentional and behavioral difficulties are referred to the Unit.

### 3.2 Sampling

Once access and permission to use the Unit as a site was granted, the researchers together with the head psychiatrist set out criteria for selection of participants. This included identification of potential participants by psychiatrists during their follow-up consultations; outpatients between the ages of 13 and 16 years, ability to speak English fluently; adolescents who have been diagnosed with ADHD for over six months. Although 14 potential participants were invited only five accepted the invitation. There was one female and four males with age range of 13-14 years. Racially, there were three blacks, one white and one Indian.

### 3.3 Procedure

The sample was obtained in two ways. The first method required the psychiatrists to identify potential participants during their follow-up consultations by using a screening form, which comprised brief demographic areas, namely, name, age, date of birth, grade, gender, mainstream or non-mainstream school. In addition, the following criteria were applied: participants were to be outpatients diagnosed with ADHD, between 13 and 16 years of age and be able to speak English fluently. The psychiatrists identified potential participants and attained informed consent from the parent for the researchers to make telephonic contact with them. In this way, the researchers obtained permission from the parents themselves and were able to explain the research study in detail. The second method used to obtain the sample required the head psychiatrist along with the psychiatric nursing staff to scan the log book patient diagnosed with ADHD. Adolescents who had attended within the previous three to four months were short-listed. Since it is customary for the nurses to make follow up telephone calls to patients, it was decided that in the spirit of these follow up calls, the nurses would use this as an opportunity to contact the parent of the adolescents that were short-listed. The nurses were instructed to stress to the parents that involvement in the study was not part of their treatment and that the study was completely voluntary. There were difficulties with incorrect and nonexistent telephone numbers. However despite this, the second method proved a far more successful way of obtaining the sample.

Once the researchers received verbal consent to contact the parents the study was explained to them in detail. The details included confidentiality, the voluntary nature of the study was time, place, length of the interview, and transport costs. The researchers also spoke with the participants to establish an initial connection.

### 3.4 Data collection

Data was collected via semi-structured interviews which lasted approximately an hour. The interviews were audio recorded. The focus of the interviews was on how the participants perceive the disorder, their competence and their relationships. Amongst the questions that were asked included, "why do you think you come here to this Unit in Baragwanath hospital? How is your behavior in the classroom? How do you understand your behavior? How easy or difficult is it for you to concentrate and learn in classroom? What is it like for you when you have to listen in class and take down the instructions? How do you get on with your peers and educators? At school have you been teased or involved in a fight because of the way you behave? If so, explain how. Are you taking medication for your concentration? If so, how do you find the medication? What do you think a concentration problem is? Is it difficult for you to sit still in your seat? Please explain. What does it feel like to have this problem? Please explain what you think of yourself? Do you ever find that your mind wanders at home like it does at school? Explain how. What are the things that have helped you with your concentration? What do you think you need to help yourself? How is your scholastic performance?

## 4. Data analysis

The audio-taped interviews were transcribed verbatim. This is a consistent method of representing verbal data in written form and it included emphasis of words, significant pauses, interruptions and overlaps in speech. The textual data underwent classical thematic content analysis (Krippendorf, 1980) to facilitate the authors' understanding of the interviews, through the interpretation of emerging themes. Thematic content analysis allows for systematic examination of the data to record the relative incidences of themes (Welman & Kruger, 2001). Common and recurrent themes in the text were identified by creating a list of categories that reflected the major themes. The identified themes were then coded and categorized (Neuman, 2006), and this was followed by analysis and interpretation of the participants' experiences (McMillan & Schumacher, 2001).

### 4.1 Ensuring trustworthiness of the data

In order to ensure the trustworthiness of the analyzed data, certain checks were put in place to verify the data and the analysis (Fade, 2003). The first author, a highly trained and experienced academic in conducting qualitative research oversaw implementation of the study. Trustworthiness was further enhanced by peer examination (the supervisor reviewed the work at all stages).

### 4.2 Reflexivity

Before the results are presented, the authors discuss their reflections, observations and experiences during the data collection. What proved to be productive in the interview were the ways in which the second author's tone and intonations were used. This facilitated the probing method, as questions needed to be non-invasive and non judgmental in order to create a safe and trustworthy climate. Eye contact, reflecting on participants' responses, as well as nodding were other crucial mechanisms used to show the participants that they were being heard and understood.

The non-verbal behavior in interview should not be ignored as they can subtly reveal more than the words are able to do. For example, the participants displayed behavior that depicted typical characteristics of ADHD. One participant frequently fiddled and played with the strings of his tracksuit top. Another, on three occasions during the interview was distracted by the tape recorder and took moments to see himself whether or not the tape was functioning. Another two often played with their fingers and rubbed their hands together. It is important to remember that these young people, who had been diagnosed with ADHD, were enduring approximately a 60 minute interview that seemed to be a challenge to them.

Emotions were also displayed non-verbally. At times, the voices of participants became very soft when they were discussing peer difficulties and poor scholastic performance. Their soft-spoken responses displayed their vulnerability. One female participant expressed her feelings through tears. Her responses in text do not appear to be emotional, however during the interview tears streamed down her face.

## 5. Results and discussion

This section presents the results of the study concurrently with the discussion. Based on thematic content analysis, four themes emerged from the data analysis, namely, problems; belief in medication; emotional distress and retribution; and feelings of worth.

### 5.1 Problems

All the participants reported having behavioral and cognitive problems which had adverse effects on their learning. These problems were largely experienced within the context of the classroom, as demonstrated by the following excerpts.

> *"I used to talk in class; I wouldn't get my work done… it was difficult to sit still, sometimes I used to stand up, watch the window and think"*
>
> *"You need to finish your schoolwork…you play around, you don't do your schoolwork that you are supposed to do…"*
>
> *"…I fiddle; I don't sit still… I move around… I walk around"*
>
> *"Sometimes when the teacher talks, you don't concentrate and you look at somebody else…you stare at that person for long"*

These excerpts appear to be consistent with Barkley's (2000) description of ADHD as a problem of behavioral disinhibition and consistent lack of persistence to sustain attention. It appears from these results that the ADHD impacted adversely on the participants' learning as they struggled to sit still, in order to be receptive to learning, as well as to complete the given tasks. Not only were these adolescents disruptive to their own learning, but to other learners as well, due to their inability remain seated. The current findings also mirror the results conducted in western countries in which the theme 'problems' emerged as the dominant theme and it comprised learning, cognitive and behavioral problems (Kendall et al., 2003).

### 5.2 Belief in medication

A powerful and dominant theme that emerged from this study was also the belief in medication. All the adolescents in the present study believed that taking medication was beneficial to them, and appeared to view medication and their experiences of ADHD as inextricably intertwined. Despite other treatment modalities available, they reported only about pills. They made a strong association between taking medication and their behavior and performance at school. The general trend was that medication improves their concentration in the classroom, inhibits their urge to talk excessively or give up from their seats and control their aggressive tendencies.

> *"When I take the tablets I don't talk a lot in class. They help me out a lot...it helps me in class to do my work and finish my work"*
>
> *"I feel I can concentrate better...'cos my grade seven marks were high"*
>
> *"I'm okay now. Sometimes I used to like get ten out of twenty and now I get twenty out of twenty for spelling. For maths I used to (sic) get nine out of hundred"*
>
> *"Like if I am on the tablet then I would sit still and time would go so fast and when I look again the day is finished...well, before I was on the tablet I didn't have Ritalin and then I would never get my work finished"*
>
> *"My term papers were very bad and since I have been on it (Ritalin), my papers have been good. In my old school all my exams used to be below 30% and this one that I just got back recently, I got like a very good mark. Information used to go in one ear and come out the other, when now it just sits there"*
>
> *"...because like in Grade 4, when I did not take my medication, my teacher used to see fast that I didn't take it. It helped me really it did help me 'cos now I can sit nicely like this..."*
>
> *"I feel happier now"*

There was a belief from the participants that the pills assisted them in their learning and as a result their scholastic performance improved. These results are consistent with previous empirical study in which the pills were seen to have positive outcomes (Kendall et al., 2003).

Medication was perceived to be helpful in controlling ADHD symptoms. Regarding interventions, their management of the disorder, all the participants perceived medication to be the primary intervention. All confirmed that the tablets they were taking were called Ritalin. They all felt their way of supporting themselves and managing their disorder was by remembering to take their medication. This was their way of taking responsibility. They all knew at what time of day they should take their medication. Furthermore, their responses indicated that they had been assisted by their teachers over the years. As they literature states, this is the one type of medication where teachers are happy to remind and assist their students in taking their medication (Rains & Scahill, 2004). When the participants were asked what they thought helped them the most, one responded, *"the tablets of course"*. Another response to this question was, *"here, the people at the hospital that give me the tablets"*.

Although the responses were largely pro-medication, there were some ambivalent feelings towards taking medication. For instance, one participant made it clear that while the medication helped her, she had the experience of being teased about taking tablets and she subsequently perceived herself to be imperfect because of this. When she made these comments in the interview she was emotional and tearful.

*"I just tell them (learners) you don't understand. They don't understand. That's why I got (sic) problems with concentration. But now – my teacher say (sic) maybe I will stop to take medication because I'm okay now. Because now, you know the last time my pills were finished, I didn't do anything like stand up. I just sit. So I talk to myself, I can do this without taking my pills'. Like the nurse in the school, she gives us (sic) our pills. Like I asked her, 'can I stop to take pill?' she told me I must talk to my teacher first. There is (sic) some of my friends who stopped medication 'cos they are fine".*

### 5.3 Emotional distress and retribution

It was interesting to note how subtly feelings of exclusion, embarrassment, disappointment and hurt were presented in the interviews. In each case, the tone of the interview changed when delicate material was conveyed. One participant did not so much speak about her feelings but displayed them non-verbally. She was able to speak with confidence and put her thoughts across but she was tearful throughout the interview.

*"Ja, like at school, it is not nice to see that my reports are bad. Like we doing (sic) a test and the teacher reads out marks and it's bad"*

*"It feels bad, like…it sucks …not nice".*

*"I'm in the remedial class. Let's just say when I failed a test, I felt sad"*

These findings corroborate Dumas and Pelletier's (1999) study, whereby it was found that preadolescents with ADHD had lower perceptions of self-worth than their counterparts in terms of scholastic competence, social acceptance and behavioral conduct. Scanlon (2006) concurs that learners with ADHD develop negative self-perceptions because of negative connotations made about their abilities when compared to other learners who do not have ADHD. Given their already compromised self-esteem, it seems critical that educators avoid actions that may be embarrassing to the learners, since actions like that may not only alienate them but also lead them to develop negativity towards school and eventual drop out of school. Elbaum and Vaughn (2003) make a significant point that learners with ADHD and other learning difficulties usually have a low academic self-concept as a result of experiences of emotional, scholastic and social failure early in their school career. It is therefore pertinent that educators and parents are not only understanding, but that they are also nurturing and accommodative of these learners.

As a result of being teased for taking the medication and not performing well at school, there was a general feeling of the need to defend oneself. Although some participants had thoughts of retaliating, the others were actually physically aggressive.

*"I had to stay back from grade 6 and now, today the grade 7 has been teasing me. Everyone says I failed and it makes me feel like hitting them"*

*They were swearing at me (sic) and then I got angry. It wasn't physical. We almost became physical. I was just sitting in class doing my work then I was just thinking, thinking, thinking, then they start swearing me..."*

*"They would tease me, say bad things or when they disrespect a woman...ja...and like Grade 7 has teased me and he's also fighting with my friends. I don't like that I have felt like hitting him. I just wanna hit them because I don't like being made fun of"*

*"Sometimes I get cross if someone even teases me. I get cross and I can't control my anger. Some people don't want to be my friend".*

*"Like sometimes, me (sic) and my friend used to get into trouble because we hit some other people. He was teasing because...we are taking pills, we mad ...whatever ...he was saying things like that. But now, no more, no more teasing. I haven't had someone teasing me in class. Sometimes they ask me why I take medication. I just tell them because of my concentration and they say 'okay', because they thought I was taking it just for fun"*

These results appear to be in line with Hutchins' (2005) view that learners with ADHD commonly experience humiliation, embarrassment and high levels of stress as educators and parents often expose their weaknesses. It appears from these excerpts that these learners may be experiencing double victimization. In addition to experiencing symptoms of ADHD which impairs their learning, they are also teased and ridiculed. As a result of such feelings, they retaliate by hitting other learners. Since they are wearing the label of being 'disruptive' they may be perceived by their significant others (educators and parents) as instigators of trouble. It is not surprising then that adolescents with ADHD were found to have fewer friends, lower self-esteem and poorer psychosocial adjustment than those without ADHD (Houghton, 2006).

In this study two participants spoke about being mean to the other learners. Their disinhibited behavior and anger seem to have acted as a defense to protect their vulnerability. The following accounts demonstrate how feelings of inadequacy, hurt and pain can at times lead to bullying.

*"I used to be very mean. I used to always hurt kids smaller than me"*
*"I feel like hitting the guy's face in, but I've gotta (sic) control myself because I don't want to get myself kicked out of school"*

Brown (2005) notes that if ADHD is not well managed through skills training and medication, conduct disorder and oppositional defiant disorder can develop during the adolescent years. It is therefore crucial that adequate and appropriate treatment is provided.

### 5.4 Feelings of worth

Despite the terrible feelings and difficulties they experienced as a result of having ADHD, these learners spoke positively of themselves. Whether or not these extracts are inflated perceptions of the self, they reflect the need to feel worthy and competent.

*"I'm good, I'm intelligent and jolly"*

*"I would say I'm cheerful, happy, and just okay"*

*"I don't struggle in class, it's just okay, and it's not hard"*

*"I am okay now, no more teasing and I am able to achieve academically"*

Since the symptoms of ADHD must be present before the age 7 years for a diagnosis to be made, it can be deduced that for most of their school years they receive attention for bad behavior. If this is the case, who will reframe their identity positively? They will. Krueger and Kendall (2001) are of the opinion that children with ADHD are likely to have exaggerated self-perceptions and these perceptions tend to serve as a self-protective mechanism to ward-off feelings of inadequacy and helplessness. Although one cannot conclude from these excerpts whether the participants were overestimating their competence and exaggerating their sense of well-being, it is encouraging to realize that they too would prefer to be valued and appreciated. This reminds us then that by being disruptive or even aggressive towards other learners, it may be a cry for help. It is therefore upon the adults to psycho-educate them about alternative adaptive ways of managing their behavior, without getting into trouble.

## 6. Limitations of the study

The research was qualitative in nature and the size and character of the sample do not allow for generalization of the research findings to the population as a whole. The sample selected in this study was sourced from a specific medical context. A variety of participants selected from support groups, schools or those who are patient in private practice may have added an interesting dimension to the experiences and self-perceptions of adolescents with ADHD.

The study did not manage to delve into the family life of adolescents with ADHD. The school setting appeared to pre-dominate the interviews. To gain a fuller understanding of the associated problem of ADHD, insight into life at home would have been a valuable component to the study. Given that this study is the first to be conducted within the African continent, it should therefore be seen as an exploratory study.

## 7. Implications of the study

The implications of this study emphasize the existence of ADHD in the everyday life of those who have it. The feeling and perceptions of the participants bring the 'human element' to this disorder. Their vulnerability, the emotional distress, the need to be acknowledged and recognized highlight the reality of this disorder. The description of their behavior, the account of their pain, and the struggles with scholastic performance all point to the fact that ADHD is not outgrown during adolescence. It is hoped that the more ADHD is really understood, the more effort can be made on the part of the parents, educators and professionals to alleviate some of the difficulties of the problem. Knowledge of the experiences of learners with ADHD will not only enable educators to appreciate and accommodate these learners in their classrooms by adapting their teaching programmes and methods, but will also enable them to use the strengths of these learners as point of departure in educating them (Holz & Lessing, 2002).

As reported in literature review, for a child or adolescent to be diagnosed with ADHD according to the DSM-IVR, some impairment of concentration must be present both at home and at school. The authors found that the participants did not perceive concentration to be a problem at home. Their responses indicated that they felt less challenged by their difficulties in their home environment.

It is interesting to note that at no point in the interviews did the participant use the label 'ADHD' or Attention Deficit Disorder (ADD) to describe their difficulties. Their descriptions however, show that they understand where their difficulties lie. Furthermore, at no point during the interview did they use the word 'attention', 'hyperactive' 'impulsive' or 'distractible'. These terms do not appear to be a part of their vocabulary or thinking even though they are used not only in the medical, educational and psychological field, but also in popular culture.

Poor scholastic performance is commonly linked with ADHD (Venter, 2009) - that of low self-esteem affects scholastic achievement and emotional adjustment. In turn, poor scholastic performance affects self-esteem (Desai, 2003). What is evident from these results is that the participants felt acutely sensitive about their ability to achieve. Their comments about failure, performing at school and their tone in the interviews all indicated feelings of disappointment, exclusion and inadequacy.

The desire to be validated, worthy and competent is not exclusive to the individual with ADHD. It is something which all human beings search for. But for the adolescent with ADHD, the need is heightened. Their school years are known to be the worst time of their lives as they are placed in an environment which is discouraging, especially by their peers (Pooley, 2002).

The participants placed importance on the need to feel worthy and competent. They initially acknowledged their pain and difficulties. They were also able to find ways to feel a sense of competence and worthiness. They tended to build a 'happy' picture for themselves regardless of their negative experiences. Furthermore, the participants' comments about needing to feel competent emphasize the significance of acknowledging children and adolescents with ADHD for any amount of effort they may display. To fuel their feelings of worth and competence, it is necessary for their support network system to praise and recognize their strengths and talents. Recognizing their value and detecting any form of competence is a means of supporting them (Hallowell & Ratey, 1994). This is confirmed by another study, which revealed that children with ADHD had fewer adjustment problems when they perceived that they had social support (Demeray & Elliot, 2001).

## 8. Conclusion

This study aimed to explore and examine the experiences, feelings and self-perceptions of adolescents who have ADHD. The sample was selected from Chris Hani Baragwanath's Child and Adolescent Psychiatric Unit. The findings and the interpretation indicate that the adolescents have insight into their difficulties. Descriptions of their behaviors and difficulties match the commonly known symptoms of ADHD. It was found that the adolescents use the term concentration problem as an umbrella term to define their difficulties. Regarding the perception around treatment and intervention, the adolescents support the use of medication. The adolescents however, had little to say about other forms of intervention. Another important finding is the participants' experiences of being teased. This appears to have a detrimental effect on their self-esteem. Their self-esteem is also severely impacted by their difficulties to perform adequately at school and scholastic performance proves to be one of major problem the adolescents' experience. Due to these difficulties, it was found that learners with ADHD have a greater need to feel competent and valued.

The findings of the study reveal that the participants do not outgrow ADHD once they reach adolescence nor are their associated problems resolved. The themes explicitly show that ADHD is not benign and affects the day-to-day experiences of the adolescent. For the participants, the school environment is the place where their difficulties and challenges come to the fore. From scholastic performance to peer relationships, ADHD has had an impact on their well-being. Several participants spoke about their vulnerabilities and pain and they also indicated their need to feel competent and worthy. Finally, the participants displayed no qualms about taking medication and to some extent, believe that it has been a cure for their difficulties.

## 9. Acknowledgements

The authors would like to thank Sue Thompson who provided expert guidance in the conceptualization and implementation of the study.

## 10. References

Aase, H, Meyer, A & Sagvolden, T. (2006). Moment to Moment dynamics of ADHD behavior in South African children. *Behavioral and Brain Function*, 2:11 doi:10.1186/1744-9081-2-11

Adewuya, A. O. & Famuyiwa, O.O (2007). Attention deficit hyperactivity disorder among Nigerian primary school children Prevalence and co-morbid condition. *European Child Adolescent Psychiatry*, 16, 10-15.

Amen, D.G. (2001). *Healing ADHD*. New York: Berkley Books.

American Psychiatric Association (1994). *Diagnostic and statistical manual of mental disorders: DSM-IV*. (4th ed.) Washington D.C.

Barkley, R. (1990). *Attention Deficit Hyperactivity Disorder: A Handbook for Diagnosis and treatment*. New York: The Guilford Press.

Barkley, R A (1998) *Attention Deficit Hyperactivity Disorder: A Handbook for Diagnosis and treatment*. (2nd Ed.) New York: The Guilford Press.

Barkley, R.A. (2000). *Taking charge of ADHD. Revised edition*. New York: Gilford Press.

Biederman, J., Willens, T.E. & Spencer, T. J. (2002) Attention Deficit/Hyperactivity Disorder across the Lifespan. *Annual Review of Medicine*, 53, 113-132.

Bird, H.R. (2002). *The diagnostic classification, epidemiology and cress cultural validity of ADHD, in Attention Deficit/Hyperactivity Disorder: State of the Science: best practices*. Kingston, NJ, Civic Research institute.

Bradley-Klug, K. & Grier, J. (2000). Adolescents and their Families: Coping with ADHD. *School Psychology Quarter*, 15(4), 480-486.

Brown, T.E. (2005). *Attention Deficit Disorder: the unfocused mind in children and adults*. New Haven: Yale University Press.

Bryant, P. & Coleman, A. (1994). (Eds.) *Developmental Psychology*. (pp.54-62). New York: Longman.

Conners C. K. & Jett, J.L. (2001). *Attention Deficit Hyperactivity Disorder*. Kansas City: Compact Clinicals.

Demeray, M.K. & Elliot, S.N. (2001). Perceived social support by children with characteristics of attention deficit hyperactivity disorder. *School Psychology Quarterly*, 16 (1), 68-91.

Desai, A. (2003). *The impact of learning in a second language and academic competence on learners' self-esteem: a study of grade four girls from 2 schools in Gauteng*. Unpublished master's thesis, University of the Witwatersrand, South Africa.

Dumas, D. & Pelletier, L. (1999). Perception in hyperactive children, Maternal Child Nursing, `24, 12-19.

DuPaul, G.J., Power, T. J., Anastopoulos, A. D., Reid, R., McGoey, K., & Ikeda, M. (1997a). Teacher ratings of ADHD symptoms: Factor structure and normative data. *Psychological Assessment, 9*, 436-444.

DuPaul, G.J. & Stoner, G.R.A. (1994). *ADHD in the schools: Assessment and intervention strategies*. New York: Guilford.

Elbaum, B. & Vaughn, S. (2003). Self-concept and students with learning disabilities. In H.L. et al. (Eds.). *Handbook of learning disabilities*. New York: The Guildford Press.

Fade, S.A. (2003). Communicating and judging the quality of qualitative research: the need for a new language. *Journal of Human Nutrition Dietetics*, 16, 139–149.

Goldman, L.S., Genel, M., Bezman, R.J. & Slanetz, P.J. (1998). Diagnosis and treatment of Attention Deficit/Hyperactivity Disorder in children and adolescents. *Journal of the American Medical Association*, 278(14), 1100-1107.

Goldstein, S. & Goldstein, M. (1998). *Managing ADHD: A Guide for Practitioners*. New York: John Wiley & Sons Inc.

Gross, M. A. (1997). *The ADD brain diagnosis treatment and science of Attention Deficit Disorder (ADD/ADHD) in adults, teenagers and children*. New York: Nova Science

Hallowell, E.M, & Ratey, J.J (1994) Answers to Distraction. New York: Bantam Books.

Holz, T. & Lessing, A. (2002). Reflections on Attention Deficit Hyperactivity Disorder (ADHD) in an inclusive system. *Perspectives in Education*, 20(3), 103-110.

Houghton, S. (2006). Advances in ADHD research through the lifespan: common themes and implications. *International Journal of Disability, Development and Education*, 53(2), 263-272.

Hutchins, P. (2005). *Helping children with learning disabilities and ADHD: working together, from inclusion to belonging congress*. Nedbank Sandton Auditorium and Conference Centre, Sandton, 12 May.

Kendall, J. (2000). Outlasting disruption: The process of reinvestment in families ADHD children. *Qualitative Health Research*, 8(6), 839-857.

Kendall, J., Hatton, D., Beckett, A., & Leo, M (2003).Children's Accounts of Attention Deficit Hyperactivity Disorder. *Advances in Nursing Sciences*, 26(2), 114- 130.

Krueger, M. & Kendall, J. (2001). Descriptions of self: An exploratory study of adolescents with ADHD, *Journal of Child and Adolescent Psychiatric Nursing*, 14(2), 61-73.

Krippendorff, K. (1980). *Content Analysis: An introduction to its methodology*. Newbury Park, CA: Sage.

Laning, H., Roake, N., Glynis, & Horning, G. (2003). *Life Soweto Style*. Cape Town: Struik Publishers.

Lambert, N. & Hartsough, C. (1998). Prospective study of tobacco smoking and substance dependencies among samples of ADHD and non-ADHD participants, *Journal of Learning Disabilities*, 31, 533-544.

Louw, C., Oswald, M.M. & Perold, M.D. (2009). General practitioners familiarity, attitudes and practices with regards to Attention Deficit Hyperactivity Disorder (ADHD) in Children and Adults, *South African Family Practice*, 51, 152-157.

Mahomedy, Z., van der Westhuizen, D., van der Linde, M.J. & Coetsee, J. (2007). Persistence of Attention Deficit/Hyperactivity Disorder into adulthood: A study conducted on parents of children diagnosed with Attention Deficit/Hyperactivity Disorder, *South African Psychiatry review*, 10, 93-98.

Mash, E.J. & Wolfe D.A. (2005). Abnormal Child Psychology. 2nd edition. Belmont: Wadsworth.

McMillan, J. H., & Schumacher, S. (2001). *Research in education: A conceptual introduction* (5th Ed.). New York, NY: Longman.

Mercugliano, M. (1999). What is ADHD? *Pediatric Clinics of North America*, 46, 831-843.

Meyer, A., Eilertsen, D.E., Sundet, J.M., Tshifularo, J. & Sagvolden, T. (2004). Cross culture similarities in ADHD - like behavior amongst South African primary school children, *South African Journal of Psychology*, 34(1) 122-138.

Nahlik, J. (2004). Issues in Diagnosis of Attention Deficit /Hyperactivity Disorder in Adolescents. *Clinical Pediatrics*, 43(1), 122-138.

Nauert, R. (2009). *Perception of ADHD behavior may be placebo-induced*. June 16 2011, from http://psychcentral.com/news/2009/06/40/perception-of-adhd-behavior-may-be-placebo-induced/6809.html.

Neuman, W.L. (2006). *Social research methods: qualitative and quantitative approaches*, 6th edition. Boston: Pearson Education.

Perold, M, Louw, C. & Kleynhans, S. (2010). Primary school teachers' knowledge and misperceptions of attention deficit hyperactivity disorder (ADHD). *South African Journal of Education*, 30, 457-473.

Polanczk, G., de Lima, M.S, Horta, B.L, Biederman, J. & Rohde, L.A. (2007). The worldwide prevalence of ADHD: A Systemic review and metaregression analysis, *American Journal of Psychiatry*, 164(6), 942-948.

Pooley, S. (2002). Names don't count. A guide to ADHD for parents. ADHD Support Group for South Africa.

Rains, A. & Scahill, L. (2004). New long-acting stimulants in children with ADHD. *Journal of Child and Adolescent Psychiatric Nursing*, 17(4), 177-180.

Rhee, S.H., Waldman, I.D., Hay, D.A. & Levy, F. (1999). Sex differences in genetic and environmental influences on DSM -III-R Attention - Deficit Hyperactivity Disorder, *Journal of Abnormal Psychology*, 108(1), 24-41.

Robin, A.L. (2002). Lifestyle issues. In Goldstein, S. & Ellison, A.T. (Eds.), *Clinician's guide to adult ADHD: Assessment and intervention*. San Diego: Academic Press.

Scanlon, D. (2006). Learning disabilities and attention deficits. In Thies, K.M. & Travers, J.F. (Eds.). *Handbook of Human Development for Health Care Professional*. Sudbury: Jones and Bartlett Publishers.

Schroeder, C.S. & Gordon, B.N, (2002). *Assessment and Treatment of Childhood Problems* (2nd ed.). London: The Guilford Press.

Seabi, J. (2010). Foundation phase educators' perceptions of Attention Deficit Hyperactivity Disorder at a mainstream primary school. *Journal of South African Higher Education, 24*(4), 616-629.

Timimi, S. & Taylor, E. (2004). ADHD is best understood as a cultural construct, *British Journal of Psychiatry*, 184, 8-9.

Van der Westhuizen, A (2010). Attention deficit Hyperactivity disorder (ADHD), *South African Pharmaceutical Journal*, 10-20.

Vogel, W. & Holford, L. (1999). Child psychiatry in Johannesburg, South Africa. A descriptive account at two clinics in 1997, *European Child & Adolescent Psychiatry*, 8, 181-188.

Welman, J.C. & Kruger, S.J. (2001). *Research Methodology* (2nd ed.) Oxford University Press: Cape Town.

# 12

# Attention Deficit Hyperactivity Disorder and Males in the Juvenile Justice System

Robert Eme
*American School of Professional Psychology, Argosy University, Schaumburg Campus*
*USA*

## 1. Introduction

Attention Deficit Hyperactivity Disorder (ADHD), a neurodevelopmental disorder of self-control/self-regulation characterized by developmentally inappropriate and impairing levels of inattention and hyperactivity-impulsivity (Barkley, 2006), is a major risk factor in the developmental sequence that results in serious antisocial behavior. Consequently, it is widely prevalent in the juvenile justice system and should be included in every mental health screening that should take place when a juvenile becomes involved with the system as well as treatment planning. The objective of this chapter is to present the empirical support for these contentions for male juvenile offenders. The focus is restricted to males since, although the correlates of antisocial behavior in male and females have typically been found to be the same (Lahey et al., 2006; Odgers et al., 2008), the literature on females is slight compared to that of males and thus problematic with regard to the reliability of the findings.

## 2. ADHD and increased risk for severe antisocial behavior

Severe antisocial behavior is legally designated, delinquency, when it refers to behavior that violates criminal law and psychiatrically designated, conduct disorder (CD), when it refers to a repetitive and persistent pattern of behavior in which the basic rights of others or major age-appropriate societal norms are violated (Farrington, 2009). There is strong empirical support establishing a robust association between ADHD and increased risk for the development of severe antisocial behavior defined either legally or psychiatrically.

With regard to legally defined antisocial behavior, clinical studies of children diagnosed with ADHD and followed-up into adolescence and early adulthood have consistently found higher rates of delinquent behavior. Mannuzza et al. (1989) compared arrest rates in 103 adolescent ADHD subjects and 100 normal controls (mean age of 18 years for both groups). Significantly more hyperactive than control subjects had been arrested for any offense (39% versus 20%), convicted for any offense (28% versus 11%), arrested for a felony offense (25% versus 7%), and incarcerated (9% versus 1%). Satterfield et al. (1982) reported an 8-year follow- up (mean age of 17 years) of 110 ADHD adolescents. The percentage of these subjects arrested at least once for a serious offense in the lower, middle, and upper socioeconomic classes was 58%, 36%, and 52% compared with 11%, 9%, and 2% for the controls. Barkley et al. (2004) compared the arrest rates of 147 ADHD and 73 control children (mean age 20-21). Those with ADHD had a greater rate of official arrests for misdemeanor (24% versus 11%) and for felony (27% versus 11%) offenses. Molina et al. (2009) in the landmark clinical Multisite Multimodal Treatment Study of children with ADHD (MTA) which had the largest clinical sample to date of children with ADHD (n=436), reported that at ages 13-18 approximately 25% to 30% of the youths were in the spectrum of clinically serious antisocial behavior, 26.8% were arrested at least once by 8 years, and 30% had engaged in moderately serious to serious delinquent behavior according to youth or parent report. All these outcomes were significantly higher than the control group, and the increased risk for serious antisocial behavior also characterized the group of children who had received 14 months of intensive state-of-the-art combination of behavior therapy and medication management. Moreover, these findings most probably underestimated the magnitude of the risk of the development of serious antisocial behavior associated with ADHD for two reasons. First, the children who were lost to follow-up tended to come from demographically disadvantaged families and thus at greater risk for antisocial behavior than those who remained in the study (Molina et al., 2009). Second, because at the time of the follow up most children had not entered late adolescence, the peak period for antisocial behavior (Tremblay, 2010), the difference in risk between the two groups again is most probably an underestimate. Lastly, Bussing et al. (2010) in a large epidemiological study, screened a school district sample of 1,625 students aged 5 to 11 years for ADHD. The 94 youth diagnosed with ADHD were followed up 8 years later and compared to matched case controls. Children with ADHD were three times more likely to be involved with the juvenile justice system based upon parental report than case controls (19% vs. 6%). Furthermore, because as with the Molina et al. study (2009), most children had not entered the peak period for antisocial behavior at the 8 year follow-up, the difference in risk for involvement in the juvenile justice system between the two groups is most probably an underestimate.

With regard to psychiatrically defined antisocial behaviour, numerous clinical studies of child onset cases of CD, which represent more severe cases of CD and thus are most similar to delinquent youth, there is the striking finding that the vast majority of males with CD are comorbid for ADHD (Beauchaine, Hinshaw, & Pang, 2010; Frick & Moffitt, 2010; Lahey, Loeber, Burke, & Applegate, 2005; Klein et al., 1997; McMahon & Frick, 2007).

In summary, given the very strong association between ADHD and the risk of developing severe antisocial behavior, it would seem that its role as a causal risk factor for this development is secure. However, there is significant controversy over the role ADHD plays in the development of CD and delinquency because of the failure to take into account oppositional defiant behavior (ODD) [Burke, Waldman, & Lahey, 2010]. Namely, although developmental pathway theories of serious antisocial behavior commonly posit ADHD as the first step in the sequence, followed by ODD, with more severe CD behaviors emerging later (Frick & Marsee, 2006; Waschbusch, 2002), ADHD is rejected as a direct developmental precursor to CD because it does not consistently predict CD if prior ODD is taken into account (Burke, Waldman, & Lahey, 2010; Loeber, Burke, & Pardini, 2009). This issue will be addressed by first examining the role of ODD in the development of CD.

## 3. ODD as a developmental precursor to CD

The essential features of ODD are a recurrent pattern of negativistic, defiant, disobedient, and hostile behavior toward authority figures, which leads to impairment (APA, 2000). Its role as a developmental precursor to CD has been well documented (Frick & Marsee, 2006; Lahey, Loeber, Burke, & Applegate, 2005; Moffitt et al., 2008). Moreover, it is now understood that far from being simply a benign, milder form of CD, ODD plays a key role in the development of CD and is one of the strongest predictors of the onset of CD and of the course of CD symptoms over time (Loeber, Burke, & Pardini, 2009). And, although the majority of children with ODD do not go on to develop CD (Loeber, Burke, & Pardini, 2009), if childhood onset CD develops, it is almost always preceded developmentally by ODD (Burke, Waldman, & Lahey, 2010). This development is likely to occur when social contexts (e.g., family and peer environment) increase rather than decrease the antisocial propensity of ADHD/ODD (Dick, 2011; Lahey & Waldman, 2008; Meier et al., 2008; Murray & Farrington, 2010). In addition, there is emerging evidence that there are subdimensions of ODD symptomatology that are not equally associated with the risk of developing CD (Pardini, Moffitt, & Frick, 2010; Rowe et al,. 2010). The symptoms that index a negative affect dimension predict internalizing problems whereas oppositional symptoms such as *often argues with adults, often actively defies or refuses to comply with adult's requests or rules*, which index a "headstrong" dimension, predict CD (Rowe et al,. 2010; Stringaris & Goodman, 2009). The "headstrong" dimension of ODD has been found to be associated with ADHD (Stringaris & Goodman, 2009) and thus provides an apt segue to consider the role of ADHD in the development of ODD.

## 4. ADHD and ODD

ODD is the most common comorbid condition of ADHD in children and studies of clinic-referred ADHD children find that between 54% and 67% will meet criteria for a diagnosis of ODD by 7 years of age or later (Barkley, 2006). This comorbidity is best explained by the core impairments of behavioral and emotional impulsivity in ADHD (Barkley, 2006, 2010a, 2010b). Although it is without question that behavioral impulsivity is a core impairment in ADHD Combined Type (ADHD-C), and is receiving even more emphasis in the proposed change for DSM 5 to add four more symptoms of impulsivity (Castellanos, 2010), it is only recently that emotional impulsivity/dysregulation has also been recognized as a core ADHD impairment (Barkley, 2010a, 2010b). These twin impairments commonly result in symptoms such as irritability, impatience, anger, low frustration threshold, and reactive aggression (Barkley, 2010a; Frick & Viding, 2009) which greatly increase the risk for coercive, oppositional interchanges (Barkley, 2006; Burns & Walsh, 2002; Lahey & Waldham, 2008; van Lier, van der Ende, Koot, & Verhulst, 2007). Indeed, it is estimated that a typical child with ADHD has an astonishing half a million of these negative interchanges each year (Pelham & Fabiano, 2008), thereby adding impressive support to Barkley's judgment (2010a) that having ADHD-C virtually creates a borderline case of ODD in children.

Finally, perhaps the most persuasive evidence that because ADHD is a disorder of impaired self-control/self-regulation it increases the risk for ODD and thereby increases the risk for CD, comes from the most recent findings of the landmark Dunedin Multidisciplinary Health and Development study (Moffitt, Arseneault, Belsky, et al.. 2011). This longitudinal study, which has followed a complete birth cohort of 1,037 children from birth to age 32, found that self control assessed at age 3 predicted criminal offending at age 32. When the sample was segmented into the highest and lowest fifths on preschool self control, the lowest fifth had much higher crime conviction rates than the highest fifth: 43% vs. 13%. Since the measures used to assess self-control were essentially measures of the core features of the behavioral and emotional impulsivity that characterize ADHD-C (hyperactivity, impulsivity, inattention, lack of persistence, impulsive aggression, low frustration tolerance) in effect what the study found was that preschool children with many symptoms of ADHD (although they were not formally diagnosed as such in the study) were at high risk for criminality compared to those with good self-control. Thus impressive support has been added to the consensus that "Self-control theory is one of the most thoroughly researched and cited theories of deviance, delinquency, and crime" (Vaszosnyi & Huang, 2010, p.245) and that impulsivity is the most crucial variable that predicts antisocial behavior (Farrington, 2009).

In conclusion, there is mountainous evidence that ADHD indirectly increases the risk for early onset CD by greatly increasing the risk for ODD. This developmental sequence explains the robust consensus that "ADHD symptoms appear to be nearly ubiquitous among clinic-referred boys who meet diagnostic criteria for CD at 7-12 years of age" (Lahey, Loeber, Burke, & Applegate, 2005, p. 396).

## 5. CD Comorbid with ADHD

Although ADHD is a major risk factor for the development of serious antisocial behavior such as CD or delinquency, these ominous developments only occur in a minority of those with ADHD/ODD (as previously discussed) when social contexts (e.g., family and peer environment) increase rather than decrease the antisocial propensity. If this development occurs however, juveniles who are comorbid for CD and ADHD display a more pernicious form of antisocial behavior than those with a single disorder in terms of a greater range, severity, and persistence of antisocial activity and greater academic impairment (Barkley, 2006;). For example, although over half of the cases with child onset CD do not persist into adult life, such persistence is much more likely if the CD is comorbid with ADHD (Rutter, 2011). Similarly, delinquents with ADHD, in comparison to delinquents without ADHD, have earlier and more frequent involvement with the criminal justice system and a greater frequency and severity of aggression when incarcerated (Young & Goodwin, 2010).

## 6. Prevalence of ADHD in the juvenile criminal justice system

Since the vast majority of males with severe, child onset CD are comorbid for ADHD (Beauchaine, Hinshaw, & Pang, 2010), it logically follows that ADHD should be widely prevalent in the juvenile justice system. This indeed is the case as the following two types of research studies conducted in the United States have found.

First, prevalence of ADHD has been estimated from diagnoses made prior to juveniles involvement with justice system. Individual education plans (IEP) indicate that 28% of juveniles enter the correctional system with a prior diagnosis of ADHD (Tudisco, 2006). Note that IEP data underestimates the true prevalence of ADHD since it does not include those juveniles who either have a prior diagnosis of ADHD that is not included in the IEP or are undiagnosed.

Secondly, prevalence of ADHD is estimated through assessments when the juvenile is in the justice system using a variety of self-report methods. Earlier narrative reviews of these studies reported estimates ranging from 20% to 72% (American Academy of Pediatrics, 2001; Vermeiren, 2003). A more recent narrative review of 17 studies of incarcerated, detained, & secured juvenile populations from 1990 to 2003 reported the following rates for males: a) age 13 and younger -12.5%, b) ages 14-15 - 20.9%, c) ages 16 and older - 13.8% (Templin et al., 2006). A meta-analytic review of 13 studies of adolescents in detention or correction facilities reported an average rate of 11.7% for males (Fazel, Doll, & Langstrom, 2008).

Lastly, Washburn et al. (2007) reported rates of 17.1% for males in 1,829 randomly selected youths in the Cook County temporary juvenile detention center in Chicago, IL. This study will assume special importance in interpreting the various foregoing estimates because it is the single best study to date as its large sample of 1,829 randomly selected youths is most representative of the entire country for the following reasons: a) most juvenile detainees nationwide live in or are detained in urban areas, b) Cook County is racially/ethnically diverse, c) the juvenile system in Cook County is typical of most other states, and d) the gender, age, and offense distributions in the detention center are similar to those detained nationwide (Washburn et al., 2007).

Given the wide range of estimates, what is the best estimate and why? Before the question can be answered, it must be noted that all of the studies which estimate ADHD prevalence in the correctional system have two serious limitations. First, the ADHD diagnosis is based exclusively on self-report. Since accurate diagnosis requires information from significant others such as parents and teachers, the failure to include such information results in under diagnosis (Barkley, 2006; Barkley, Murphy, & Fischer, 2008). For example, Young et al. (2010) conducted a study to determine the most reliable source of information of ADHD symptoms by comparing rating scales scored by 54 male delinquents who were detained in a high risk care home and their teachers with psychiatric diagnosis of ADHD made by a professional clinical assessment. Sensitivity rates were 33% for delinquent self report compared to 67% for teacher report. Additional clear and convincing evidence that substantial underdiagnosis has indeed occurred in studies of the prevalence of ADHD delinquents comes from the previously discussed numerous clinical studies of juveniles who, having been diagnosed CD, were also assessed for ADHD using information from parents and teachers. In these studies, as previously discussed, the vast majority of those with CD were also diagnosed with ADHD. If this finding is applied to the 53% of males and females in detention or incarceration who have CD (Washburn et al., 2007) [note that 53% is an underestimate as it is based on self-report], then the vast majority of the 53% with CD would also presumably be diagnosed with ADHD if they were properly evaluated. Second, all studies that have assessed for ADHD in adolescents in all settings suffer from another limitation that results in an underestimation of true prevalence. Namely, these measures are based upon developmental manifestations of ADHD in children which become increasing irrelevant to the subsequent developmental manifestations of ADHD at subsequent ages (Barkley, 2010c; Barkley, Murphy, & Fischer, 2008; Kessler et al., 2010; Malone et al., 2010). For example, child criteria emphasize overt hyperactivity which diminishes with age and many of the criteria that index symptoms of inattention in childhood are less relevant to subsequent manifestations of inattention (Barkley, Murphy, & Fischer, 2008; Kessler et al., 2010).

In summary, if approximately 28% of those who enter the juvenile justice system have a prior formal diagnosis of ADHD, if the vast majority of 53% plus of juveniles in the system with CD would also be diagnosed with ADHD (if they were properly assessed), if the reported rates of 17.1% for males in the best study to date (Washburn et al., 2007) are an underestimate because of the inadequacies of assessment for ADHD, then a conservative prevalence estimate of 25% would seem to be reasonable.

### 6.1 International studies

Studies in countries other than the United States (Canada, Sweden, Finland, Norway, and the United Kingdom) reported an ADHD prevalence rate averaging 45% among juveniles in the correctional system (Young & Goodwin, 2010; Young et al., 2011). These findings provide strong confirmation that the 25% prevalence rate in the United States is indeed a conservative estimate. In addition, the 20% difference in the prevalence rates may in fact represent a real difference that cannot simply be attributed to methodological variance. For example, it may be that more of the delinquency in the United States is due to cultural pathologies, e.g. gang affiliation (Farrington, 2009), whereas proportionately more of the delinquency in other countries is due to individual pathologies such as ADHD.

### 6.2 Conclusion

There is no doubt that the international scientific literature has found that ADHD is widely prevalent in the juvenile justice system in virtually all countries that have been studied. Furthermore, it is quite striking that this prevalence is vastly disproportionate to what would be expected based upon population rates. Though population prevalence rates vary widely because of methodological rather than true differences (National Institute for Health and Clinical Excellence, 2009), a systematic review of international prevalence studies by Polanczyk et al. (2007), after correcting for methodological variance, found only minor differences in different countries and reported an average rate of about around 5.3%. Since this rate blends male and female rates and since the male rate in epidemiological studies is about 2 to 3 times higher (Barkley, 2006), a rough international estimate would yield a male rate of 7% and female rate of 3%. Thus the prevalence rate of ADHD among male juvenile offenders is at least 4 times more common than what would be expected from the population rate. This widespread prevalence of ADHD among male juvenile offenders makes it imperative that every mental health screening that should take place when a juvenile becomes involved with the justice system should screen for ADHD. The 25% to 45% of the individuals who can be expected to be diagnosed with ADHD receive competent treatment for this disorder, as well as for whatever other disorders might be comorbid with the ADHD.

## 7. Screening for ADHD

The awareness that the mental health needs of many youth in the juvenile justice system are unmet has led to the development of systematic mental health screening tools in an effort to better identify and respond to these youth (Skowyra & Cocozza, 2007). However, to judge from the most widely used screening tool in juvenile justice, the *Massachusetts Youth Screening Instrument, 2nd Version (MAYSI-2)* [Skowyra & Cocozza, 2007], these tools fail to adequately screen for ADHD. Hence the mental health needs of the 25% to 45% of youth in the justice system with ADHD are not met. If these needs are to be met, the following screening recommendations for ADHD are suggested.

### 7.1 Screening recommendations

Mental health screening is a relatively brief process performed by non-clinical staff using a standardized screening instrument which may result in a follow-up comprehensive clinical assessment for youth whose scores indicate they may have mental health problems (Williams, 2007). One way in which screening for ADHD could be incorporated into a standard mental health screening would be to add an ADHD screening module to the most widely used screening tool in juvenile justice, the *Massachusetts Youth Screening Instrument, 2nd Version (MAYSI-2)*. Such a module should include the following elements.

1. *Carefully review existing documentation.*

Examine all of the juvenile's existing documentation to see if it includes a prior diagnosis of ADHD in childhood. If it does, a referral for a more comprehensive confirmatory evaluation is warranted since most children diagnosed with ADHD continue to have the disorder into adolescence (Barkley, 2006).

2. *Screen for self-reported attention problems.*

Attention problems are the single most important symptom cluster for identifying ADHD beyond childhood (Barkley, 2010c; Barkley, Murphy & Fischer, 2008). The following questions have been found to be the most sensitive indicators of these problems (Barkley, 2010c; Barkley, Murphy & Fischer, 2008; Keller et al., 2010).

- *Are you easily distracted?*
- *Do you have difficulty sustaining attention?*
- *Do you have difficulty prioritizing work?*
- *Do you have trouble planning ahead?*
- *Do you have difficulty completing tasks on time?*

If the individual indicates that these problems occur often or very often, this is indicative of ADHD (Barkley, 2010c) and warrants referral for a more comprehensive confirmatory evaluation.

## 8. Treatment

The following treatment recommendations are based upon the consensus statement of the United Kingdom ADHD network and criminal justice agencies for the management of juvenile offenders with ADHD (Young et al., 2011). For whichever stage an offender is at (police, court, probation, parole, prison) a protocol should to be established for insuring that intervention meets the needs of offenders with ADHD. This protocol should have three broad aspects:

1. Pharmacological treatment to alleviate ADHD symptoms.
2. Psychological treatment to improve self-control, reduce antisocial attitudes and behaviors, and treat comorbid disorders.
3. Integration of care pathways.

### 8.1 Pharmacological treatment

An extensive literature has clearly documented the robust short-term efficacy (2-8 weeks) of stimulant treatment for core ADHD symptoms of inattention, impulsivity and hyperactivity for school-age Caucasian boys (Biederman & Spencer, 2008; Connor, 2006). These studies have also found that the core symptoms of ADHD in children with ADHD and ODD/CD respond as well to stimulant treatment as symptoms in children with ADHD alone (Pliszka, 2009). The effect size of the response of stimulants relative to placebo (response to which is generally low, e.g., 13%) is close to 1.0 thereby making stimulants among the most efficacious medications in all of health care, rivaling the antibiotics in this regard (Pliszka, 2009).

Stimulants also have similarly robust effects on oppositional/defiant behaviors and overt aggression (Connor 2006; Connor et al., 2002; Pliszka, 2009), especially when an optimal medication regimen is combined with behavior therapy (Blader et al., 2010). Perhaps the best indication of the efficacy of treating children comorbid for ADHD/ODD comes from the largest randomized control study to date, the Multimodal Treatment Study of ADHD, in which 40% of the participants had ODD or CD (Smith, Barkley, Shapiro, 2006). Fourteen months of combined stimulant and behavior treatment resulted in a success rate of 68% (defined as an excellent response such that functioning was in the normal range, i.e., no or almost no symptoms of ADHD or ODD) in contrast to a 25% success rate for the community comparison group (Swanson et al., 2001).

#### 8.1.1 Potential for abuse of stimulant medication

The delivery of stimulant medication in a prison setting should not be a problem since these settings typically already run medication-based programs for controlled drugs and successfully adhere to policies that reduce the chances of mismanagement (Young et al., 2011).

The potential for abuse of stimulant medication that is delivered in other settings can be reduced in three ways. First, the most conservative approach would be to use a non-stimulant medication such as atomoxetine. This approach however suffers from the limitation that it is less effective than the stimulants (Faraone, 2009; Young et al., 2011). Second, stimulant medications can be used that are less vulnerable to abuse. Prodrug or long acting formulations can be used, as they are less easily manipulated than immediate release formulations (Kollins, 2008). Also delivery systems such as the crush resistant shell of Concerta or a methylphenidate skin patch can be used (Mariani & Levin, 2007). Third, there should be careful monitoring for signs of possible abuse or diversion such as missed appointments, repeated requests for higher doses or a pattern of 'lost' prescriptions (Kollins, 2008).

### 8.2 Psychological treatments

Stimulant treatment should always form part of a comprehensive treatment plan that includes psychological, behavioral, and educational interventions (Young et al., 2011). Specific programs have been developed that integrate stimulant treatment with non-pharmacological treatments for ADHD that enhance the efficacy of medication and also treat comorbid problems (Young & Amarasinghe, 2010).[1] Furthermore, these combined treatments can improve adherence to whatever offender programs the correctional system offers (Young et al., 2011).

### 8.3 Integration of care pathways

It is critical to establish a continuous, integrated care pathway that follows the offender from initial police contact through to eventual disposition (Young et al., 2011). An adequate knowledge of ADHD throughout the whole care pathway is critical in helping ADHD offenders rehabilitate. Most importantly, there must be community services to support ADHD offenders in implementing this continuity of care.

## 9. Conclusion

ADHD is a major risk factor for the development of male juvenile antisocial behavior. This risk results in a conservatively estimated international ADHD prevalence rate of 25% to 45% of males in juvenile correctional systems. Therefore, given this widespread prevalence, it follows that an adequate understanding of ADHD must exist at every stage of the offender pathway from initial police contact to eventual disposition. Screening for ADHD must be implemented in all criminal justice services. Having identified ADHD in an offender, appropriate treatment needs to be implemented in an integrated fashion throughout the care pathway.

## 10. References

American Academy of Pediatrics (2001). Health care for children and adolescents in the juvenile correctional care system. *Pediatrics, 107,* 799-803.

American Psychiatric Association. (2000). *Diagnostic and statistical manual of mental disorders* (Text rev.). Author, ISBN-10: 0890420254, Washington DC.

Barkley, R. (2006). *Attention-deficit hyperactivity disorder* (3rd ed.).Guilford Press, ISBN-10: 9781593852108, New York.

Barkley, R. (2010a). Deficient emotional self-regulation is a core symptom of ADHD. *Journal of ADHD and Related Disorders, 1,* 5-37.

Barkley, R. (2010b). The ADHD report: why emotional impulsiveness should be a central feature of ADHD. *ADHD Report 18,* 1-4.

Barkley, R. (2010c). *Taking charge of adult ADHD.* Guilford Press, ISBN-10: 1606233386, New York.

Barkley, R., Fischer, M., Smallish, L., & Fletcher, K. (2004). Young adult follow-up of hyperactive children: Antisocial activities and drug use. *Journal of Child Psychology and Psychiatry, 45,* 195-211.

[1] As it is beyond the scope of this chapter to discuss the specific programs, this article should be consulted as it provides an excellent review which details recommended interventions, the level of evidence on which the recommendation is based, and a brief description of techniques employed.

Barkley, R., Murphy, K., & Fischer, M. (2008). *ADHD in adults.* Guilford Press, ISBN-10: 9780763765644, New York.

Beauchaine, T., Hinshaw, S., Pang, K. (2010). Comorbidity of attention-deficit/hyperactivity disorder and early-onset conduct disorder: biological, environmental, and developmental mechanisms. *Clinical Psychology: Science and Practice,17,* 327-336.

Blader, J., Pliszka, S. Jensen, P., Schooler, N. & Kafantaris, V. (2010). Stimulant-responsive and stimulant-refractory aggressive behavior among children with ADHD. *Pediatrics, 126,* 796-806.

Burke, J., Waldman, I., & Lahey, B. (2010). Predictive validity of childhood oppositional defiant disorder and conduct disorder: Implications for the DSM-V. *Journal of Abnormal Psychology, 119,* 739-751.

Burns, G., & Walsh, J. (2002). The influence of ADHD on the development of oppositional defiant disorder symptoms in a 2-year longitudinal study. *Journal of Abnormal Child Psychology, 30,* 245-256.

Bussing, R., Mason, D., Bell, L., Porter, P. & Garvan, C. (2010). Adolescent outcomes of childhood attention-deficit/hyperactivity disorder in a diverse community sample. *Journal of the American Academy of Child and Adolescent Psychiatry, 49,* 595-605.

Castellanos, F. (2010). DSM-5: Options being considered for ADHD. *American Psychiatric Asssociation,* http://www.dsm5.org/ProposedRevisions/Pages/proposedrevision.aspx?rid=383#

Connor, D. (2006). Stimulants. In R. Barkley (Ed.), *Attention-Deficit Hyperactivity Disorder* (3rd ed., pp. 658-677). New York: Guilford Press.

Connor, D., Glatt, S., Lopez, I., Jackson, D., & Melloni, R. (2002). Psychopharmacology and aggression. I: A meta-analysis of stimulant effects on overt/covert aggression-related behaviors of ADHD. *Journal of the American Academy of Child & Adolescent Psychiatry, 41,* 253-261.

Dick, D. (2011). Gene-environment interaction in psychological traits and disorders. *Annual review of clinical psychology, 7,* 383-409.

Eme, R. (2010). Male life-course-persistent antisocial behavior: the most important pediatric mental health problem. *Archives of Pediatric and Adolescent Medicine, 164,* 186-187.

Faraone, S. (2009). Using meta-analysis to compare the efficacy of medications for Attention-Deficity/Hyperactivity Disorder in youths. *P & T, 34,* 678-694.

Farrington, D. (2009). Conduct disorder, aggression, and delinquency. In R. Lerner & L. Steinberg (Eds.), *Handbook of adolescent psychology, individual bases of adolescent development, Vol.1* (pp. 683-721). Wiley, ISBN-10:0470149213, New York.

Fazel, S., Doll, H., & Langstrom, N. (2008). Mental disorders among adolescents in juvenile detention and correctional facilities: a systematic review of metaregression analysis of 25 surveys. *Journal of the American Academy of Child and Adolescent Psychiatry, 47,* 1010-1019.

Frick, P., & Marsee, M. (2006). Psychopathy and developmental pathways to antisocial behavior in youth. In C. Patrick (Ed.), *Handbook of psychopathy* (pp. 353-374). Guilford Press, ISBN-10: 1593855915, New York.

Frick P. & Moffitt T. (2010) A proposal to the DSM-V Childhood Disorders and the ADHD and Disruptive Behavior Disorders Work Groups to include a specifier to the diagnosis of conduct disorder based on the presence of callous-unemotional Traits. *American Psychiatric Association, 1–36.*Retrieved from. http://www.dsm5.org/ProposedRevisions/Pages/proposedrevision.aspx?rid=424 # aspx?rid=424#

Frick, P., & Viding, E. (2009). Antisocial behavior from a developmental psychopathology perspective. *Development and Psychopathology, 21,* 1111-1131.

Kessler, R., Green, J., Adler, L., Barkley, R., Chatterji, S., Faraone, S., et al. (2010). Structure and diagnosis of adult ADHD. *Archives of General Psychiatry, 67,* 1168-1178.

Klein, R., Abikoff, H., Klass, E., Ganeles, D., Seese, I., & Pollack, S. (1997). Clinical efficacy of methylphenidate in conduct disorder with and without attention deficit hyperactivity disorder. *Archives of General Psychiatry, 54,* 1073–1080.

Kollins, S. (2008). ADHD, substance abuse disorders and stimulant treatment: Current literature and treatment guideline. *Journal of Attention Disorders, 12,* 115-125.

Lahey, B., Loeber, R., Burke, J., & Applegate, B. (2005). Predicting future antisocial personality disorder in males from a clinical assessment in childhood. *Journal of Consulting and Clinical Psychology, 73,* 389–399.

Lahey, B., Van Hulle, C., Waldman, I., Rodgers, J., D'Onofrio, B., Pedlow, S., et al. (2006). Testing descriptive hypotheses regarding sex differences in the development of conduct problems and delinquency. *Journal of Abnormal Child Psychology, 34,* 737-755.

Lahey, B., & Waldman, W. (2008). Personality dispositions and the development of violence and conduct problems. In D. Flannery, A. Vazsonyi & I. Waldman (Eds.), *The Cambridge handbook of violent behavior and aggression* (pp. 260–287). Cambridge University Press, ISBN-10: 052160785X, New York.

Loeber R., Burke, J., & Pardini, D. (2009). Perspectives on oppositional defiant disorder, conduct disorder, and psychopathic features. *Journal of Child Psychology and Psychiatry, 50,* 133-142.

Malone, P., Van Eck, K., Flory, K., & Lamin, D. (2010). A mixed-model approach to linking ADHD to adolescent onset of illicit drug use. *Developmental Psychology, 46,* 1543-1555.

Mannuzza, S., Klein, R., Konig, P., & Giampino, T. (1989). Hyperactive boys almost grown up. *Archives of General Psychiatry, 46,* 1073-1079.

Mariani, J., & Levin, F. (2007). Treatment strategies for co-occurring ADHD and substance use disorders. *The American Journal of Addictions, 16,* 45-56.

McMahon, R. & Frick, P. (2007). Conduct and oppositional disorders. In E. Mash & R. Barkley (Eds.), *Assessment of childhood disorders (4th ed.),*(pp. 131-183). Guilford Press, ISBN-10: 9781606236154, New York.

Meier, M., Slutske, W., Arndt, S., & Cadoret, R. (2008). Impulsive and callous traits are more strongly associated with delinquency behavior in higher risk neighborhoods among boys and girls. *Journal of Abnormal Psychology, 117*, 377-385.

Moffitt, T. Arseneault, L., Belsky, D., Dickson, N., Hancox, R. Harrington, H., & Caspi, A. (2011). A gradient of childhood self-control predicts health, wealth, and public safety. *Proceedings of the National Academy of Sciences of the United States of America, 108*, 2693-2698.

Moffitt, T., Arseneault, L., Jaffee, S., Kim-Cohen, J., Koenen, K, Odgers, C., & Viding, E. (2008). Research review: DSM-V conduct disorder: research needs for an evidence base. *Journal of Child Psychology and Psychiatry, 49*, 3-33.

Molina, B., Hinshaw, S., Swanson, J., Arnold, L.,Vitiello, B., Jensen, P., et al. (2009). The MTA at 8 Years: Prospective Follow-up of Children Treated for Combined-Type ADHD in a Multisite Study. *Journal of the American Academy of Child and Adolescent Psychiatry, 48*, 484-55.

Murray, J., & Farrington, D. (2010). Risk factors for conduct disorder and delinquency: key findings from longitudinal studies. *Canadian Journal of Psychiatry, 55*, 633-642.

National Institute for Health and Clinical Excellence (2009). *Attention deficit hyperactivity disorder: Diagnosis and management of ADHD in children, young people, and adults. NCE Clinical Guideline 72*. London.

Odgers, C., Moffitt, T., Boradbent, J., Dickson, N., Hancox, R., Harrington, H., et al. (2008). Female and male antisocial trajectories: from childhood origins to adult development. *Development and Psychopathology, 20*, 673-716.

Pardini, D., Moffitt, T. & Frick, P. (2010). Building an evidence base for DSM-5 conceptualizations of oppositional defiant disorder and conduct disorder: Introduction to the special section. *Journal of Abnormal Psychology, 119*, 683-688.

Pelham, W., & Fabiano, G. (2008). Evidence-based psychosocial treatments for Attention-Deficit/Hyperactivity Disorder. *Journal of Child and Adolescent Psychology, 37*, 184-214.

Pliszka, S. (2009). *Treating ADHD and comorbid disorders: psychosocial and psychopharmacological interventions*. New York: The Guilford Press.

Polanczyk, G., Silva de Lima, M., Horta, B., et al. (2007) The worldwide prevalence of ADHD: a systematic review and metaregression analysis. *American Journal of Psychiatry, 164*, 942–948.

Rowe, R. Costello, E., Angold, A., Copeland, W., & Maughan, B. (2010).Developmental pathways in oppositional defiant disorder and conduct disorder. *Journal of Abnormal Psychology, 119*, 726-738.

Rutter, M. (2011). Research review: Child psychiatric diagnosis and classification: concepts, findings, challenges and potential. *Journal of Child Psychology and Psychiatry, 52*, 647-660.

Satterfield, J., Hoppe, C., & Schell, A. (1982). A prospective study of delinquency of 110 adolescent boys with attention deficit disorder and 88 normal adolescent boys. *American Journal of Psychiatry, 139*, 795-798.

Skowyra, K., & Cocozza, J. (2007). *Mental health screening within juvenile justice: the next frontier*. New York: National Center for Mental Health and Juvenile Justice.

Smith, B., Barkley, R., & Shapiro, C. (2006). Combined child therapies. In R. Barkley (Ed.), *Attention-Deficit Hyperactivity Disorder* (3rd ed., pp. 678-691). New York: Guilford Press.

Snyder, H. & Sickmund, M. (2006) Juvenile offenders and victims: 2006 National report. *National Center for Juvenile Justice, March,* 234-235.

Stringaris, A. & Goodman, R. (2009). Three dimensions of oppositionality in youth. *Journal of Child Psychology and Psychiatry, 50,* 216-223.

Swanson, J., Kraemer, H., Hinshaw, S., Arnold, L., Conners, C. Abikoff, H., & Wu, M. (2001). Clinical relevance of the primary findings of the MTA: success rates based on severity of ADHD and ODD symptoms at the end of treatment. *Journal of the American Academy of Child & Adolescent Psychiatry, 40,* 168-179.

Templin, L., Abram, K., McClelland, G., Mericle, A., Dulcan, M., & Washburn, J. (2006). Psychiatric disorders in detention. *Juvenile Justice Bulletin, April,* 1-16.

Tremblay, R. (2010). Developmental origins of disruptive behavior problems: the 'original sin' hypothesis, epigenetics and their consequences for prevention. *Journal of Child Psychology and Psychiatry, 51,* 341-367.

Tudisco, R. (2006, October). Impact of ADHD on the juvenile/criminal justice system. *Presentation at the 18th Annual International Conference on AD/HD,* Chicago, IL.

Van Lier P., Van der Ende J., Koot H., & Verhulst F. (2007). Which better predicts conduct problems? The relationship of trajectories of conduct problems with ODD and ADHD symptoms from childhood and into adolescence. *Journal of Child Psychology and Psychiatry. 48,*601-608.

Vaszosnyi, A., & Huang, L. (2010). Where self-control theory comes from: on the development of self-control and its relationship to deviance over time. *Developmental Psychology, 46,* 245-257.

Vermeiren, R. (2003). Psychopathology and delinquency in adolescents: A descriptive and developmental perspective. *Clinical Psychology Review, 23,* 277-318.

Waschbusch, D. (2002). A meta-analytic examination of co-morbid hyperactive-attention impulsive problems and conduct problems. *Psychological Bulletin, 128,* 118–150.

Washburn, J., Romero, E., Welty, L., Abran, K., Templin, L., McClelland, G., et al. (2007). Development of antisocial personality disorder in detained youths: the predictive value of mental disorders. *Journal of Consulting and Clinical Psychology, 75,* 221-231.

Young, S., Adamou, M., Bolea, B., Gudjonsson, G., Muller, U., Pitts, M., et al. (2011). The identification and management of ADHD within the criminal justice system: a consensus statement from the UK adult ADHD network and criminal justice agencies. *BMC Psychiatry, 11,* 1-14.

Young, S., & Amarasinghe, M. (2010). Practitioner review: non-pharmacological treatments for ADHD: a lifespan approach. *Journal of Child Psychology and Psychiatry, 51,* 116-113.

Young, S., & Goodwin, E. (2010). Attention-deficit/hyperactivity disorder in persistent criminal offenders: the need for specialist treatment programs. *Expert Review of Neurotherapeutics, 10,*1497-1500.

Young, S., Gudjonsson, G., Misch, P., Collins, P., Carter, P., Redfern, J. et al. (2010) Prevalence of ADHD symptoms among youth in a secure facility: the consistency and accuracy of self- and informant-report ratings. *Journal of Forensic Psychiatry and Psychology, 21*, 238-246.

Williams, V. (2007). *Mental health screening within juvenile justice: the next frontier*. New York: National Center for Mental Health and Juvenile Justice.